Lecture Notes in Mathematics

Edited by A. Dold and B. Eckmann

T0220013

764

Bhama Srinivasan

Representations of Finite Chevalley Groups

A Survey

Springer-Verlag
Berlin Heidelberg New York 1979

Author

Bhama Srinivasan
Department of Mathematics
Clark University
Worcester, Massachusetts 01610
USA

AMS Subject Classifications (1970): 20C15, 20G40

ISBN 3-540-09716-3 Springer-Verlag Berlin Heidelberg New York
ISBN 0-387-09716-3 Springer-Verlag New York Heidelberg Berlin

Library of Congress Cataloging in Publication Data
Srinivasan, Bhama, 1935-
Representations of finite Chevalley groups.
(Lecture notes in mathematics; 764)
Bibliography: p.
Includes index.
1. Chevalley groups. 2. Finite groups. 3. Representations of groups.
I. Title. II. Series: Lecture notes in mathematics (Berlin); 764.
QA3.L28 no. 764 [QA171] 510'.8s [512'.22] 79-26526
ISBN 0-387-09716-3

Printing and binding: Beltz Offsetdruck, Hemsbach/Bergstr.
2141/3140-543210

To women in mathematics:

May their tribe increase

INTRODUCTION

The aim of these notes is to give a survey of the main developments in the theory of "ordinary", ie. characteristic representations of finite Chevalley groups which have occurred in recent years. In the year 1969-70 a seminar on finite groups arising from algebraic groups was held at the Institute for Advanced study. In this seminar T.A. Springer gave some lectures on Harish-Chandra's philosophy of cusp forms, which had at that time been applied also to these finite groups. Springer then stated the so called "Macdonald conjectures" which predicted that there would be families of representations of these groups parametrized by the characters of the various "maximal tori". An important breakthrough came in 1976 when Lusztig and Deligne in their famous paper [18] published a proof of these conjectures, by constructing virtual representations of the groups on the ℓ-adic cohomology of certain varieties.

An outline of the contents is as follows. Chapter I is a review of the main results that we need on the "absolute theory" of reductive algebraic groups over an algebraically closed field. In Chapter II and in the rest of the notes we consider the situation where G is a connected reductive group defined over $\mathbb{F}_q$, and where F is a Frobenius endomorphism of G. The group G^F of fixed points under F is the finite group whose representation theory will be studied. The classification of the maximal tori of G^F is described, leading us to the problem of constructing a family of virtual representations of

G^F corresponding to each torus. In Chapter III this is done in the easiest case, ie. in the case of the "split torus", leading to the principal series representations. In Chapter IV the theory of Harish-Chandra is described and this brings out the importance of constructing cuspidal (discrete series) representations.

Chapter V is perhaps the raison d'etre for these notes. In recent years I have detected a growing dissatisfaction among finite group theorists about assuming the existence and main properties of ℓ-adic cohomology on an axiomatic basis. I have therefore endeavored in this chapter, starting from a brief review of the classical theory of sheaves on a topological space and of sheaf cohomology, to give an idea of how ℓ-adic cohomology groups are constructed and to give a feeling for their properties by pointing out classical analogues when possible. It is hoped that this chapter will be of independent interest.

Chapter VI contains the main results in the paper of Lusztig and Deligne. If T is an F-stable maximal torus of G, a virtual representation $R_T^G(\theta)$ of G^F is constructed corresponding to each character θ of T^F (ie. homomorphism of T^F into $\bar{\mathbb{Q}}_\ell^*$, where ℓ is different from p, the characteristic of $\mathbb{F}_q$). If θ is regular, ie. not fixed by any non-trivial element of $N(T)^F/T^F$, then $R_T^G(\theta)$ is irreducible, up to sign.

Orthogonality properties of the $R_T^G(\theta)$ are established, and their dimensions are computed. The connection between this theory and the Harish-Chandra theory is established; in particular if T is a "minisotropic torus" and θ is a regular character of T^F then $\pm R_T^G(\theta)$ is cuspidal (but not all cuspidal representations arise this way). The proof of an important result in the paper [18] ((6.3) of these notes) which leads to a reduction formula for the character values of the $R_T^G(\theta)$ has so far been inaccessible to many group theorists because of the technical machinery involved. I have described the main ideas in this proof, omitting some of the details, and to my mind this is the most interesting feature of this chapter. The rest of the material follows either the Lusztig-Deligne paper [18] or the monograph of Lusztig [48].

The determination of the explicit values of the characters of the $R_T^G(\theta)$ remains one of the main unsolved problems in the theory. The work of Springer and Kazhdan which enables us to write down the values at unipotent elements in terms of "trigonometric sums" on the Lie algebra (provided p and q are large) is described in Chapter VII. Finally in Chapter VIII I have tried to bring the material up-to-date by describing recent work of Lusztig on the classification of representations of classical groups and of "unipotent" representations for all types. This chapter also contains a section on Hecke algebras ie. centralizer algebras of the representations of G^F induced from certain representations of parabolic subgroups. These algebras arise naturally when we try to decompose these induced representations.

The notes are intended to be accessible to advanced

graduate students. A knowledge of the representation theory of finite groups to the extent of say, Parts I and II of the book by Serre [66] is assumed; some knowledge of algebraic groups is desirable but not absolutely necessary. In Chapters II through VII I have given proofs of most of the results; Chapter VIII is essentially a review of recent results but I have included some discussions of proofs. The bibliography includes mainly the papers that I have quoted in the notes. For supplementary references the reader can consult a survey article by Curtis [14].

Acknowledgments

It is a pleasure to thank the Mathematics Department of the University of Illinois, Chicago Circle, for their warm hospitality during the Fall Quarter of 1978-79, when I gave the lectures which formed a base for these notes. In particular I thank the members of my audience, especially a "hard core" consisting of Paul Fong, Noboru Ito, Cary Huffman, Mark Ronan and Stephen Smith, for their stimulating comments and for their encouragement to me to publish these notes.

An initial version of the notes was written while I was a visitor at the University of Chicago in December 1978. I thank the Mathematics Department, especially Paul Sally and Jonathan Alperin, for their hospitality during this period. At this time I had several illuminating conversations with Spencer Bloch which helped me understand the material in Chapter V and it is a pleasure to thank him for this. A preliminary draft of Chapter V was read by Michael Artin who made many valuable suggestions.

I owe a great debt to George Lusztig, who has generously shared with me his time and his ideas during the last few years. His beautiful papers have led me into new worlds whose existence I was only dimly aware of earlier. I was helped in overcoming my initial trepidation at entering these worlds by conversations with, and encouragement from, David Kazhdan.

I would also like to record here my gratitude to my colleagues Robert Kilmoyer, Edward Cline and John Kennison, who have provided me through the years with mathematical stimulation, friendship, support, and a happy family atmosphere in the Department. It is also a pleasure to thank Theresa Shusas who has single-handedly run the Department with a rare combination of efficiency and good humor; in particular I thank her for the fine job she has done of typing a part of these notes. A major part of the notes was typed by Margaret Jaquith who stepped in when time was short and did an excellent job.

Finally I thank the National Science Foundation for financial support in the form of Grant MCS-78-02184.

Bhama Srinivasan

Clark University
Worcester, MA

Table of Contents

CHAPTER I. REVIEW OF RESULTS ON ALGEBRAIC GROUPS.

The references for this chapter are [5], [32], and [40]. We will define affine varieties, introduce linear algebraic groups over an algebraically closed field, and review some of their basic properties.

We will mostly follow the notation of [40]. Let K be an algebraically closed field. An <u>affine variety</u> $X \subset K^n$ is the set of zeros in K^n of a (finite) set of polynomials in $K[x_1,x_2,\ldots,x_n]$. The Zariski topology on K^n is defined by stipulating that the closed sets are affine varieties. Then we have the induced Zariski topology on any subset X of K^n.

An affine variety X is said to be <u>irreducible</u> if X is not the union of two proper non-empty closed subsets. Any affine variety can be written as the union of a finite number of closed irreducible subsets called its components. If the affine variety X is defined by a set of polynomials which generate an ideal I of $K[x_1,x_2,\ldots,x_n]$, X is irreducible if and only if I is a prime ideal. The ring $K[X] = K[x_1,x_2,\ldots,x_n]/I$ is called the <u>affine coordinate ring</u> of X. Each element $f + I \in K[X]$ (where $f \in K[x_1,x_2,\ldots,x_n]$) gives rise to a K-valued function on X; if $(a_1,a_2,\ldots,a_n) \in X$, we map it on $f(a_1,a_2,\ldots,a_n) \in K$. So $K[X]$ is also called the algebra of polynomial functions on X. It is a finitely generated K-algebra, and X can be identified with $\mathrm{Hom}_K(K[X],K)$. If X is irreducible, the <u>dimension</u> of X is the transcendence degree of the quotient field $K(X)$ of $K[X]$, over K.

Let $X \subset K^n$, $Y \subset K^m$ be affine varieties. We say a map $\phi: X \to Y$ is a <u>morphism</u> if $\phi(a_1, a_2, \ldots, a_n) = (\psi_1(a_1, a_2, \ldots, a_n) \ldots, \psi_m(a_1, a_2, \ldots, a_n))$ where $\psi_1, \ldots, \psi_m \in K[x_1, x_2, \ldots, x_n]$. The morphism ϕ induces a <u>comorphism</u> $\phi^*: K[Y] \to K[X]$ by $\phi^*(f) = \phi \cdot f$ and ϕ^* is a K-algebra homomorphism. If X,Y are affine varities, we can define a product variety $X \times Y$ and $K[X \times Y] \cong K[X] \otimes K[Y]$.

We can define the Zariski topology on projective n-space $\mathbb{P}^n$ over K by taking closed sets to be the common zeros of a set of homogeneous polynomials over K. The closed subsets of $\mathbb{P}^n$ are called projective varieties. A <u>quasi-projective</u> variety is an open set in a projective variety.

We regard $GL(n,K)$, the group of all $n \times n$ invertible matrices over K, as being embedded in K^{n^2}. Then a group $G \subset GL(n,K)$ is called a linear algebraic group if it is the intersection with $GL(n,K)$ of a closed subset of K^{n^2}. A map $G \to H$ of <u>linear algebraic groups</u> is a morphism if it is a morphism in the sense of affine varieties and a group homomorphism.

<u>Examples of linear algebraic groups</u>.

1. $GL(n,K)$.
2. $SL(n,K)$, the group of $n \times n$ matrices with determinant 1 over K.
3. $Sp(2n,K) = \{A \in GL(2n,K) \mid {}^tA \begin{pmatrix} 0 & J \\ -J & 0 \end{pmatrix} A = \begin{pmatrix} 0 & J \\ -J & 0 \end{pmatrix}\}$, where

   where J is the $n \times n$ matrix $\begin{pmatrix} & & 1 \\ & \cdot^{\cdot^{\cdot}} & \\ 1 & & \end{pmatrix}$.

4. $O(2n+1,K) = \{A \in GL(2n+1,K) \mid {}^tA \begin{pmatrix} 1 & 0 & 0 \\ 0 & 0 & J \\ 0 & J & 0 \end{pmatrix} A = \begin{pmatrix} 1 & 0 & 0 \\ 0 & 0 & J \\ 0 & J & 0 \end{pmatrix} \}$

   where char $K \neq 2$.

   $SO(2n+1,K) = O(2n+1,K) \cap SL(2n+1,K)$.

5. $O(2n,K) = \{A \in GL(2n,K) \mid {}^tA \begin{pmatrix} 0 & J \\ J & 0 \end{pmatrix} A = \begin{pmatrix} 0 & J \\ J & 0 \end{pmatrix} \}$

   where char $K \neq 2$.

   $SO(2n,K) = O(2n,K) \cap SL(2n,K)$.

<u>Remark</u>. The groups of 4 and 5 have to be defined differently in characteristic 2. See [9], p. 8.

6. The group of diagonal matrices over K.
7. The group of upper triangular matrices over K.
8. The group of upper triangular matrices over K with entries 1 along the diagonal.
9. $G_2(K)$: A group of type G_2 over $\mathbb{C}$ can be defined as the group of automorphisms of a Cayley algebra. In analogy with this, we can define a group $G_2(K)$ over K to be generated by 7×7 matrices over K satisfying certain conditions (see e.g. [58], p. 400).

Let G be a linear algebraic group. Then there is a unique irreducible component G^o of G containing the identity element, and G^o is a normal subgroup of finite index. G is connected (in the Zariski topology) if and only if G is irreducible, and if and only if G has no closed subgroup of finite index. (See [40], 7.3.)

Example. $O(2n,K)$ is not connected, whereas $SL(2n,K)$ is connected.

Definitions. 1. G is simple if it has no closed connected non-trivial normal subgroups.

Example. $SL(n,K)$.

2. G is semisimple if it has no closed connected non-trivial normal abelian subgroups.

Examples. $SL(n,K)$, $Sp(2n,K)$, $SO(2n+1,K)$, $SO(2n,K)$.

3. G is unipotent if it is isomorphic to a closed subgroup of the group of upper triangular $n\times n$ matrices over K with entries 1 along the diagonal, for some n.

Example. The additive group of K is isomorphic to the group $\{\begin{pmatrix} 1 & a \\ 0 & 1 \end{pmatrix}\}$, $a \in K$.

4. G is reductive if it has no closed connected non-trivial normal unipotent subgroups.

Examples. $GL(n,K)$, $CSp(2n,K)$ (the group of symplectic similitudes).

5. G is a torus if it is isomorphic to a product of multiplicative groups of K.

Example. The group of $n\times n$ diagonal matrices over K.

In the famous Chevalley Seminar of 1956-58 [11] the semisimple groups over K were classified up to isomorphism. The simple ones fall into families of types A,B,C,D (classical groups) or G_2, F_4, E_6, E_7, E_8 (exceptional groups). In the notation we have used above, $SL(n,K) = A_{n-1}(K)$, $SO(2n+1,K)$

$= B_n(K)$, $SO(2n,K) = D_n(K)$, $Sp(2n,K) = C_n(K)$. (See [40], §32.)

For the rest of this chapter we assume that G is a connected reductive linear algebraic group over K and state certain properties of G.

Definition. A maximal connected solvable subgroup of G is called a Borel subgroup of G.

Proposition 1.1. (See [40], 21.3, 23.1) (i) All the maximal tori in G are conjugate in G.

(ii) All the Borel subgroups of G are conjugate.

(iii) If B is a Borel subgroup then $N_G(B) = B$.

The dimension of a maximal torus is called the rank of G. G has a maximal connected normal solvable subgroup called its radical, and the quotient is semisimple. The rank of the quotient group is called the semisimple rank of G. Let T be a maximal torus of G. Then $N(T)/T$ is a finite group called the Weyl group of T and denoted by $W(T)$. It is a finite reflection group. We have the Bruhat decomposition $G = \bigcup_{w \in W(T)} B\dot{w}B$ where $\dot{w}$ is a representative for $w \in W$ in $N(T)$, and the double cosets are disjoint. We have $N(T) \cap B = T$. (See [40], 28.3.)

A subgroup P of G is called a parabolic subgroup if it contains a Borel subgroup. A parabolic subgroup P has a Levi decomposition, ie. a semidirect product decomposition $P = LV$ where V is the maximal connected unipotent normal subgroup of P and is called the unipotent radical of P. L is reductive and is called a Levi subgroup. It is not unique, but any two Levi subgroups are conjugate. In particular, if B is a Borel subgroup we have $B = TU$ where T is a maximal torus and U is a maximal unipotent subgroup of G. (See [40], 30.2.).

For any maximal torus T let $X(T) = \mathrm{Hom}(T,K^*)$, the group of morphisms of T into K^* (ie. characters of T) and let $V = X(T) \underset{\mathbb{Z}}{\times} \mathbb{R}$. Then $W(T)$ acts on V and this is the natural representation of $W(T)$ as a reflection group. There is a subset Φ of $X(T)$ which is an abstract root system ([40], p. 229) in V, except that Φ may not generate V if G is not semisimple. The elements of Φ are called the roots of G with respect to T. We can choose a set of simple roots Δ in Φ such that every root is either a positive root, ie. a linear combination of roots in Δ with positive coefficients, or a negative root, ie. a linear combination of roots in Δ with negative coefficients. Then $\Phi = \Phi^+ \cup \Phi^-$ where $\Phi^+(\Phi^-)$ is the set of positive (negative) roots. For each $\alpha \in \Phi$ there is a T-stable connected unipotent subgroup U_α of G and isomorphisms $x_\alpha : K \to U_\alpha$ such that $tx_\alpha(u)t^{-1} = x_\alpha(\alpha(t)u)$ $(t \in T, u \in K)$. G is generated by the root subgroups U_α $(\alpha \in \Phi)$ and T. The subgroup U generated by the U_α $(\alpha \in \Phi^+)$ is a maximal unipotent subgroup. Then TU is a Borel subgroup, and in fact the choice of a Borel subgroup containing T is equivalent to the choice of a set of simple roots in Φ. The subgroup of G generated by the U_α $(\alpha \in \Phi^-)$ is called the opposite of U and is denoted by U^-. For each $w \in W(T)$ Let $U_w^- = U \cap \dot{w}U^-\dot{w}^{-1}$. Then we have the following refinement of the Bruhat decomposition. Every element x in G can be written uniquely as $x = u\dot{w}tu'$ where $u \in U$, $u' \in U_w^-$, $t \in T$. (See [40], 26.3, 27.3, 28.4.)

Remark. For an easy treatment of the Bruhat decomposition in the case of SL_n and Sp_{2n} see [80], p. 73.

Finally we mention the Jordan decomposition of elements of G. Suppose $x \in GL(V)$ where V is a finite-dimensional vector space over K. Then we say x is semisimple if x is a diagonalizable automorphism of V and that x is unipotent if all of its eigenvalues are 1. If x is an arbitrary element of $GL(V)$ we have $x = x_s x_u = x_u x_s$ where x_s is semisimple and x_u is unipotent; x_s and x_u are determined uniquely by these conditions. Now let $x \in G$. Then we have $x = su = us$ where, in any embedding $G \hookrightarrow GL(V)$, s maps onto a semisimple element and u onto a unipotent element. We call s the semisimple part of x and u the unipotent part of x. (See [40], 15.3.) A torus of G consists entirely of semisimple elements and a unipotent subgroup consists entirely of unipotent elements.

CHAPTER II. CLASSIFICATION OF TORI.

In this chapter we will assume that $K = \bar{\mathbb{F}}_p$ for some prime p.

Let X be an affine variety over K. We say X is defined over $\mathbb{F}_q \subset K$, or has an $\mathbb{F}_q$-rational structure, if X can be defined by a set of polynomials with coefficients in $\mathbb{F}_q$. If X and Y are affine varieties defined over $\mathbb{F}_q$ we can talk of a morphism $\phi : X \to Y$ being defined over $\mathbb{F}_q$; we require a set of polynomials $\psi_1, \psi_2, \ldots, \psi_m$ defining ϕ (see the definition in Chapter I) to be polynomials over $\mathbb{F}_q$.

Suppose X is defined over $\mathbb{F}_q$. Then the affine coordinate ring $A = K[X]$ has a subring A_0 which is an $\mathbb{F}_q$-algebra such that $A_0 \otimes_{\mathbb{F}_q} K = A$. We have a map $F : A \to A$ called the <u>geometric</u> Frobenius morphism and defined by $F(a_0 \otimes \lambda) = a_0^q \otimes \lambda$, and a map $\phi : A \to A$ called the <u>arithmetic</u> Frobenius morphism defined by $\phi(a_0 \otimes \lambda) = a_0 \otimes \lambda^q$ ($a_0 \in A_0$, $\lambda \in K$). Then $F\phi = \phi F$ is the map $a \to a^q$ of A. F is an algebra homomorphism and is a bijection from A onto A^q, and for each $a \in A$ there is an $n \geq 1$ such that $F^n a = a^{q^n}$. (For the details of this see the proof of 2.10.) The map ϕ is a bijection and is a semilinear ring homomorphism (ie. $\phi(\lambda a) = \lambda^q \phi(a)$ for $\lambda \in K$, $a \in A$), and for each $a \in A$ there is an $n \geq 1$ such that $\phi^n(a) = a$. Giving an $\mathbb{F}_q$-rational structure on X is equivalent to defining an $\mathbb{F}_q$-algebra $A_0 \subset A$ with $A_0 \otimes K = A$, and this in turn is equivalent to giving either a map F or ϕ with the above properties. For instance, if we are given ϕ, we define A_0 as the set of fixed points of A under ϕ. The fact that for each a we have $\phi^n(a) = a$ for some n is used to show that $A_0 \otimes K = A$, as follows. If V is the subspace of A generated by $\{a, \phi a, \ldots, \phi^{n-1}a\}$, it can be shown that $V^\phi \subset A_0$ is an $\mathbb{F}_q$-space of dimension equal to $\dim V$, and so $a = \sum a_i\lambda_i$ with $a_i \in A_0$, $\lambda_i \in K$.

Since F is an algebra homomorphism of A we have a morphism $F:X \to X$ of which $F:A \to A$ is the comorphism. We call F the <u>Frobenius morphism</u> of X associated with the $\mathbb{F}_q$-rational structure on X. In fact, since we are regarding X as being embedded in K^n for some n, we can think of F as just the restriction to X of the standard Frobenius map of K^n which takes each "coordinate" x_i to $x_i{}^q$. In particular, the set X^F of fixed points of X under F is finite.

<u>Proposition 2.1</u>. If $g:X \to X$ is an automorphism of X of finite order which commutes with $F:X \to X$, then Fg is the Frobenius map of X for some $\mathbb{F}_q$-rational structure.

<u>Proof</u>. From the map $F:X \to X$ we get maps $F:A \to A$ and $\phi:A \to A$. We also have a map $g:A \to A$ of finite order. It is sufficient to show that ϕg has the required properties to define an $\mathbb{F}_q$-rational structure on A, and thus we have to show that given $a \in A$, $(\phi g)^n(a) = a$ for some n. But this is clear since ϕ and g commute and g is of finite order.

Suppose now we have a linear algebraic group G which is defined over $\mathbb{F}_q$ and let $F:G \to G$ be a Frobenius map of G compatible with the group structure, ie. F is a morphism of linear algebraic groups. The group of fixed points G^F is a finite group. The aim of these notes is to study the characteristic 0 representation theory of finite groups of the form G^F, when G is connected and reductive.

A very important tool in studying the structure of G^F is a theorem of Lang (see [5], 16.4). We give below a simple proof of this theorem due to Steinberg [84]. We first state a few definitions and results from algebraic geometry.

<u>Definition</u>. Let $\phi: X \to Y$ be a morphism of affine varities. Then ϕ is <u>finite</u> if $K[X]$ is integral over $\phi^*(K[Y])$. ϕ is dominant if X is irreducible and $\phi(X)$ is dense in Y. (See [40], pp. 30,31.)

<u>Proposition 2.2</u>. ([40], 4.2.) A finite dominant morphism of affine varieties is surjective.

<u>Proposition 2.3</u>. ([40], p. 30.) If $\phi: X \to Y$ is a dominant morphism of irreducible affine varieties and $e = \dim X - \dim Y$, then for any $y \in Y$ the dimension of any irreducible component of the fibre $\phi^{-1}(y)$ is at least e.

Now consider the algebra $A = K[G]$. Then G acts on A by left translations λ_x and right translations $\rho_x (x \in G)$ where $(\lambda_x f)(y) = f(x^{-1}y)$, $(\rho_x f)(y) = f(yx)$ $(f \in A, x, y \in G)$. An important fact is that the left or right translates under G of any $f \in A$ span a finite-dimensional K-subspace of A ([40], 8.6).

We now state Lang's Theorem.

Theorem 2.4. (Lang) If G is a connected linear algebraic group defined over $\mathbb{F}_q$ and F is the Frobenius map, then the map $\phi: x \to x(Fx)^{-1}$ of G into G is surjective.

Proof. (Steinberg) First we note that ϕ is dominant. For the fibres of ϕ are cosets of G^F and thus finite, ie. of dimension 0. Hence $\phi(G)$ has the same dimension as G by Proposition 2.3, and thus ϕ is dominant. Thus by Proposition 2.2 it is sufficient to show that ϕ is finite.

Consider $A = K[G]$. Then $A = A_0 \underset{\mathbb{F}_q}{\otimes} K$ where A_0 is an $\mathbb{F}_q$-algebra and $F(f) = f^q$ for all $f \in A_0$. It is then sufficient to show that A_0 is integral over $(F-1)A_0$. (Here, as before, we are using F also to mean the homomorphism $A \to A$ which is the comorphism of $F: G \to G$.) We note that A_0 is the algebra of $\mathbb{F}_q$-valued functions on G and hence is stable under left and right translations by G.

Let $f \in A_0$. Then if $y, z \in G$ we can write $f(yz) = \sum_i e_i(y) f_i(z)$, where $\{e_i\}$ is a (finite) basis for the space of right G-translates of f and $\{f_i\}$ is a (finite) basis for space of left G-translates of f. Then $f_j(yz) = \sum_i e_{ji}(y) f_i(z)$ where the e_{ji} are in the space of left-right translates of f, which is again finite-dimensional. Now put $y = (Fz)z^{-1}$. Then $f_j(yz) = f_j(Fz) = (Ff_j)(z)$, whereas $\sum_i e_{ji}((Fz)z^{-1}) f_i(z)$ $= \sum_i ((F-1)e_{ji} \cdot f_i)(z)$. This shows that $Ff_j = \sum_i (F-1)e_{ji} \cdot f_i$

for each j. We also have $Ff_j = f_j^q$, and thus $(F-1)A_0[\{f_j\}]$ is finitely generated as an $(F-1)A$-module. Furthermore each f_j lies in $(F-1)A_0[\{f_j\}]$ by the above expression for Ff_j, and hence so does f. This shows that A_0 is integral over $(F-1)A_0$ and proves the theorem.

We now remark that if $F:X \to X$ is a Frobenius morphism of an affine variety and if Y is a closed subvariety of X then to say that Y is defined over $\mathbb{F}_q$ is equivalent to Y being F-stable. For, Y is defined by an ideal $I \subset A$. Then $FY = Y$ if and only if $FI = I^q$. Hence if $FY = Y$ then F defines a map $A/I \to A^q/I^q$ and this leads to an F_q-rational structure on Y.

We give some examples of Frobenius morphisms of linear algebraic groups and their groups of fixed points.

Example 1. Let $G = GL(n,K)$, $F:(x_{ij}) \to (x_{ij}^q)$. Then F is called the standard Frobenius morphism and $G^F = GL(n,q)$. Similarly from $Sp(2n,K)$, $S_0(m,K)$ we get the finite groups $Sp(2n,q)$, $SO(m,q)$.

Example 2. Let $G = GL(n,K)$, and $\tilde{F}$ the "twisted Frobenius" morphism $(x_{ij}) \to (x_{ji}^q)^{-1}$. Then $\tilde{F}$ is the standard Frobenius morphism of Example 1 composed with a "graph automorphism" (see [81], p. 169) and by (2.1) this gives another $\mathbb{F}_q$-rational $G^{\tilde{F}}$ structure on G. We have $G^{\tilde{F}} \cong U(n,q)$.

Example 3. Let $G = Sp(2,K) \times Sp(2,K) \times \ldots \times Sp(2,K)$ (n copies). Let $\tilde{F}$ be the morphism of G obtained by composing the standard Frobenius morphism with the automorphism of order n cyclically permuting the factors. Then $\tilde{G}^F \cong Sp(2,q^n)$.

<u>From now on we assume for the rest of these notes, that G is a connected reductive group defined over $\mathbb{F}_q$ with a Frobenius morphism F compatible with the group structure.</u> We define a <u>torus</u> of G^F to be a subgroup of the form T^F where T is an F-stable maximal torus of G. We will now describe the classification of tori in G^F (see [7], E-20).

Let H be a closed F-stable subgroup of G.

<u>Definition</u>. The elements $x_1, x_2 \in H$ are <u>F-conjugate</u> in H if there exists an element $y \in H$ such that $x_2 = yx_1(Fy)^{-1}$.

<u>Lemma 2.5</u>. If two elements $x_1, x_2 \in H$ are F-conjugate modulo H° they are F-conjugate in H.

<u>Proof</u>. Let $x_2 = yx_1(Fy)^{-1}t$, $y \in H$, $t \in H^\circ$. Then we write $x_2 = a(Fa)^{-1}$, for some $a \in G$. This gives $a^{-1}yx_1(Fy)^{-1}(Fa) = (Fa)^{-1}t^{-1}(Fa)$. Since H° is normal in H, $(Fa)^{-1}t^{-1}(Fa) \in a^{-1}H^\circ a$, and $a^{-1}H^\circ a$ is a connected group. Thus $(Fa)^{-1}t^{-1}(Fa) = (a^{-1}t_1a)(Fa)^{-1}(Ft_1)^{-1}(Fa)$ for some $t_1 \in H^\circ$, and so $yx_1(Fy)^{-1} = t_1x_2(Ft_1)^{-1}$. Thus x_1, x_2 are F-conjugate in H.

In the same spirit, we prove

<u>Lemma 2.6</u>. Let H be connected and suppose H is normal in N. Then $(N/H)^F \simeq N^F/H^F$.

<u>Proof</u>. Let $F(Ha) = Ha$, then $a(Fa)^{-1} \in H$. Since H is connected we have $a(Fa)^{-1} = b(Fb)^{-1}$ for some $b \in H$. Then

$ba^{-1} \in Ha$ and is fixed by F.

Proposition 2.7. Let $\mathcal{S}$ be an F-stable set of subgroups of G such that G acts transitively on $\mathcal{S}$ by conjugation.

(i) There exists $S \in \mathcal{S}$ which is fixed by F.

(ii) Assume $N(S) = N$ is a closed subgroup of G. Then there is a bijection between the set of G^F- conjugacy classes of F-fixed elements of $\mathcal{S}$, the set of F-conjugacy classes of N, and the set of F-conjugacy classes of N/N^o.

Proof. Let $R \in \mathcal{S}$. Then $F(R) = a^{-1}Ra$ for some $a \in G$. Let $a = b(Fb)^{-1}$, then $b^{-1}Rb$ is F-stable. Set $b^{-1}Rb = S$. Every element of $\mathcal{S}$ can be written as $a^{-1}Sa$ for some $a \in G$. If $a^{-1}Sa$ is F-stable then $a(Fa)^{-1} \in N(S) = N$. If $b^{-1}Sb$ is also F-stable and $g^{-1}a^{-1}Sa\,g = b^{-1}Sb$ $(g \in G^F)$, then $n = agb^{-1} \in N$ and $a(Fa)^{-1} = nb(Fb)^{-1}(Fn)^{-1}$. Thus we have a well-defined map from the set $\{G^F$-conjugacy classes of F-stable elements in $\mathcal{S}\}$ to the set {F-conjugacy classes of N}, which is injective. It is also surjective, since given $n \in N$ we have $n = a(Fa)^{-1}$ for some $a \in G$ and $a^{-1}Sa$ maps onto N. Furthermore, if $n_2 = n(n_1)(Fn)^{-1}$ and $n_1 = a(Fa)^{-1}$, $n_2 = b(Fb)^{-1}$ we have $a^{-1}nb = g \in G^F$ and $g^{-1}a^{-1}Sa\,g = b^{-1}Sb$. This proves everything except the last part of (ii), which follows from Lemma 2.5.

Corollary 2.8. (i) There exists a maximal torus T and a Borel subgroup B such that $T \subset B$ and T,B are fixed by F. There is exactly one such pair up to G^F-conjugacy.

(ii) Let T be as in (i), and $N(T)/T = W(T)$. Then the G^F- conjugacy classes of F-stable maximal tori of G (and hence the conjugacy classes of tori in G^F) are in bijection with the F-conjugacy classes of $W(T)$.

Proof. (i) Take $\mathcal{S}$ to be the set of Borel subgroups of G, and use the fact that $N(B) = B$ (see 1.1). Thus all Borel subgroups of G are G^F-conjugate. Then we apply Theorem 2.7 to B, and use the fact that $N(T) \cap B = T$.
(ii) Take $\mathcal{S}$ to be the set of maximal tori of G.

From now on we denote by T_0 and B_0 a pair of subgroups chosen as in 2.8 (i). Let U_0 be the unipotent radical of B_0 so that $B_0 = T_0 U_0$. The group $W(T_0)$ is called the Weyl group of G.

Example. If $G = GL(n,K)$ and F is the standard Frobenius morphism we can take T_0 to be the group of diagonal matrices and B_0 to be the group of upper triangular matrices. Then $W(T_0) \cong S_n$ (symmetric group of degree n). F is trivial on $W(T_0)$ and the conjugacy classes of tori of G^F are in bijection with the conjugacy classes of S_n, ie. with partitions of n.

Now suppose T is any F-stable moximal torus of G. Then we can show that T^F can be thought of as being obtained from T_0^F by "twisting by an element of W", where $W = W(T_0)$. Let $\dot{w}$ be a representative for $w \in W$ in $N(T_0) = N$. Let $T = a^{-1}T_0 a$ for some $a \in G$, then $a(Fa)^{-1}$ normalizes T_0

and thus we can assume that $a(Fa)^{-1} = \dot{w}$ for some $w \in W$. (Note that w is the element of W which corresponds to T in Corollary 2.8 (ii).) We have $T^F = \{a^{-1}ta \mid t \in T_0, (Fa)^{-1}(Ft)(Fa) = a^{-1}ta\} = \{a^{-1}ta \mid t \in T_0, \dot{w}^{-1}t\dot{w} = Ft\}$. Hence $aT^Fa^{-1} = \{t \in T_0 \mid \dot{w}^{-1}t\dot{w} = Ft\}$. So T^F is isomorphic (in fact conjugate in G) to the set of fixed points in T_0 of a "twisted Frobenius", ie. $\dot{w}F$.

Next we determine the structure of $N(T)^F/T^F$.

<u>Definition</u>. If $w' \in W$, let $C'(w') = \{w \in W \mid ww'(Fw)^{-1} = w'\}$.

<u>Proposition 2.9</u>. Suppose T corresponds to $w' \in W$. Then $N(T)^F/T^F \cong C'(w')$.

Proof. Suppose $T = a^{-1}T_0a$ $(a \in G)$. Then $a(Fa)^{-1} = \dot{w}' \in N$. We have $N(T)/T = a^{-1}T_0a$ and $N(T)^F/T^F = (a^{-1}Na)^F/(a^{-1}Ta)^F$. Consider the map $(a^{-1}Na)^F \to W$ given by $a^{-1}na \to w$, where $n \in N$ corresponds to $w \in W$. Then $a^{-1}na \in (a^{-1}Na)^F$ implies that $w'^{-1}ww' = Fw$, or $w \in C'(w')$. Thus $(a^{-1}Na)^F$ maps into $C'(w')$, and the kernel is $(a^{-1}T_0a)^F$. The map is surjective by a reasoning similar to that of Lemma 2.5.

We now consider the orders of the tori T^F. We first state some facts about character groups of tori (see [11], Expose 4).

Recall (Chapter I) that if T is a maximal torus $X = X(T)$ denotes the group $\mathrm{Hom}(T,K^*)$ of rational characters of T. If T is isomorphic to a product of ℓ copies of K^* then X is a free abelian group of rank ℓ and $\mathrm{Hom}_{\mathbf{Z}}(X,K^*) \cong T$.

If A is a closed subgroup of T its annihilator A^o in X is a subgroup of X such that X/A^o has no p-torsion, for $px \in A^o$ $(\chi \in X)$ implies that $\chi(t^p) = 1$ for all $t \in A$ and so $\chi(t) = 1$ for all $t \in A$. Next, if B is a subgroup of X such that X/B has no p-torsion then $B^{oo} = B$, where $B^o = \{t \in T | \chi(t) = 1$ for all χ in $B\}$. This is seen as follows. Let $E = X/B$, then $B^o = \mathrm{Hom}(E,K^*)$ and thus it is sufficient to show that $\bigcap_{\chi \in B^o} \ker \chi = 0$. But this follows from the fact that if C is a finitely generated abelian group without p-torsion ie. $C \cong C' \times Z^m$ where C' is a finite abelian group without p-torsion, then $\hat{C} = \mathrm{Hom}(C,K^*) \cong \hat{C}' \times K^{*m}$ and $\hat{\hat{C}} = \mathrm{Hom}(\hat{C},K^*) \cong C' \times \mathbb{Z}^m \cong C$.

The order of the torus T^F is now given by the following theorem (see [7], E-22).

<u>Theorem 2.10</u>. Let $V = X(T_0) \otimes_{\mathbb{Z}} R$. Suppose the torus T corresponds to $w \in W = W(T_0)$ in the sense of (2.8)(ii). Then $|T^F| = |\det(wF-1)|$ where we regard w,F as linear transformations of V. (Note that there is a natural action of F and W on $X(T_0)$, hence on V.)

<u>Proof</u>. By Lang's theorem, $F-1$ is injective on $K[T_0]$ and hence on $X = X(T_0) \subset K(T_0)$. Furthermore, we can show that $X/(F-1)X$ has no p-torsion, as follows. Let $A = K(T_0)$, then $A = A_0 \otimes_{\mathbb{F}_q} K$ where A_0 is an $\mathbb{F}_q$-algebra. Then F takes $a \in A_0$ to a^q, and for each $a \in A$ there is an n such that

$F^n a = a^{q^n}$ (since $\sum_i \lambda_i a_i$ with $\lambda_i \in K$, and each λ_i is in a finite extension of $\mathbb{F}_q$). Since A is finitely generated there is an n such that $F^n a = a^{q^n}$ for all $a \in A$. This says that F is nilpotent on X, mod p. Thus $X/(F-1)X$ has finite order prime to p.

Now we see by the elementary divisor theorem that $|X/(F-1)X| = |\det(F-1)|$. But the annihilator $((F-1)X)^0$ of $(F-1)X$ in T_0 is T_0^F, since $(F-1)\chi(t) = t$ for all $\chi \in X$ if and only if $\chi(F(t)) = \chi(t)$ for all χ, i.e. if and only if $Ft = t$. This shows that $|T_0^F| = |\det(F-1)|$.

Now consider the torus T which corresponds to $w \in W$. Then by the remarks made earlier, T^F is isomorphic to the set of fixed points in T_0 of another Frobenius morphism ie. $\dot{w}F$. By the above arguments we then get $|T^F| = |\det(wF-1)|$, as required.

Example. $G = GL_2$, F the standard Frobenius. We have $T_0 = \{ \begin{pmatrix} \alpha & 0 \\ 0 & \beta \end{pmatrix} \}$, $\alpha, \beta \in K$. The space V has a basis $\varepsilon_1, \varepsilon_2$ where $\varepsilon_1, \varepsilon_2 \in X$ are defined by $\varepsilon_1 : \begin{pmatrix} \alpha & 0 \\ 0 & \beta \end{pmatrix} \to \alpha$, $\varepsilon_2 : \begin{pmatrix} \alpha & 0 \\ 0 & \beta \end{pmatrix} \to \beta$.

The roots are $\pm(\varepsilon_1 - \varepsilon_2)$. The non-trivial element w of $W \simeq S_2$ maps ε_1 on ε_2 and ε_2 on ε_1. We have $F : \varepsilon_1 \to q\varepsilon_1$ $\varepsilon_2 \to q\varepsilon_2$, and $wF : \varepsilon_1 \to q\varepsilon_2$, $\varepsilon_2 \to q\varepsilon_1$. Thus $|T_0^F| = (q-1)^2$, and the order of the torus T^F corresponding to w is q^2-1. More generally the order of the torus in $GL(n,q)$ corresponding to the partition $\{1^{r_1} 2^{r_2} \ldots k^{r_k}\}$ of n is $(q-1)^{r_1}(q^2-1)^{r_2}$ $\ldots(q^k-1)^{r_k}$.

Remark 2.11. If the torus T_0^F is isomorphic to a product of copies of $\mathbb{F}_q$, we say G is split over $\mathbb{F}_q$ and that the group G^F is an untwisted group.

We close this chapter with some remarks on the Jordan decomposition and the Bruhat decomposition in G^F.

If $x \in G$ then it has a Jordan decomposition as an element of G. We see easily, for example by regarding x as an element of $GL(n,K)$ for some n, that x is semisimple if and only if x is a p'-element and that x is unipotent if and only if x is a p-element.

Proposition 2.12. The finite group G^F has a Bruhat decomposition $G^F = \bigcup_{w \in W^F} B_0^F \dot{w} B_0^F$ as a disjoint union of double cosets of B_0^F. Furthermore every element x of G^F can be written uniquely as $x = u\dot{w}tu'$ for some $w \in W^F$, $u \in U_0^F$, $t \in T_0^F$, $u' \in ((U_0)_w^-)^F$.

Proof. For simplicity we will denote U_0 by U in this proposition. If $x \in G^F$ we can write x uniquely as $x = u\dot{w}tu'$ for some $w \in W$, $u \in U$, $t \in T_0$, $u' \in U_w^-$. Hence $x = unu'$ where $n \equiv \dot{w} (\text{mod } T_0)$ and $Fx = x = (Fu_1)(Fn)(Fu_2)$. Now U is F-stable since B_0 is F-stable, and U^- is F-stable since the Borel subgroup T_0U^- is G^F-conjugate to B_0. So we must have $n \equiv Fn \ (\text{mod } T_0)$, ie. $Fw = w$, ie. $w \in W^F$. Then U_w^- is F-stable, and we get $u_1 \in U^F$, $u_2 \in (U_w^-)^F$. By Lemma 2.6 every element w

of W^F has a representative $\dot{w} \in N(T_0)^F$. This shows that $G^F = \bigcup_{w \in W^F} B_0^F \dot{w} B_0^F$.

Remark. The finite simple groups which arise as composition factors of the groups G^F are the finite simple groups of Lie type, with the exception of the Ree and Suzuki groups. The latter arise as fixed points of an algebraic group under a power of a Frobenius morphism. The Lusztig-Deligne theory can be extended to them also, but for simplicity we are not considering them here.

CHAPTER III. PRINCIPAL SERIES REPRESENTATIONS.

We will now use the subgroup $T_0^F \subset B_0^F \subset G^F$ to construct complex representations of G^F. In this chapter and in Chapter IV we will denote, for any finite group H, the group of linear characters $\mathrm{Hom}(H,\mathbb{C}^*)$ by $\hat{H}$. Later on we will consider representations of G^F over $\bar{\mathbb{Q}}_\ell$, where ℓ is a prime different from p.

Let $\lambda \in \hat{T}_0^F$, and let $\tilde{\lambda} \in \hat{B}_0^F$ be the pullback of λ to $B_0^F = T_0^F U_0^F$. We induce $\tilde{\lambda}$ to G^F and obtain a representation of G^F denoted by $\mathrm{Ind}_{B_0^F}^{G^F}(\tilde{\lambda})$. The irreducible constituents of the $\mathrm{Ind}_{B_0^F}^{G^F}(\tilde{\lambda})$, where λ varies over $\hat{T}_0^F$, are called the principal series representations (or characters) of G^F.

We now recall Mackey's Theorem (see eg. [26], p. 51, or [66], p. 59).

<u>Theorem 3.1</u>. (Mackey). Suppose K_1, K_2 are subgroups of a finite group H. and suppose $H = \bigcup_x K_1 x K_2$, a union of disjoint double cosets. If ϕ_1, ϕ_2 are representations of K_1, K_2 respectively then $(\mathrm{Ind}_{K_1}^H(\phi_1), \mathrm{Ind}_{K_2}^H(\phi_2))_H$ $= \sum_x (\phi_1^x, \phi_2)_{K_1^x \cap K_2}$. In particular if $K_1 = K_2$ then $\mathrm{Ind}_{K_1}^H(\phi_1)$ is irreducible if and only if the representations ϕ_1^x and ϕ_1 of $K_1^x \cap K_1$ are disjoint.

<u>Definition</u>. $\lambda \in \hat{T}_0^F$ is regular if it is not fixed by any non-trivial element of $W(T_0)^F$.

<u>Proposition</u> 3.2. (i) $\mathrm{Ind}_{B_0^F}^{G^F}(\tilde{\lambda})$ is irreducible if and only λ is regular.

(ii) $(\mathrm{Ind}_{B_0^F}^{G^F}(\tilde{\lambda}), \mathrm{Ind}_{B_0^F}^{G^F}(\tilde{\mu}))_{G^F} = |\{w \in W(T_0)^F | \lambda^w = \mu\}|$. In particular, $(\mathrm{Ind}_{B_0^F}^{G^F}(\tilde{\lambda}), \mathrm{Ind}_{B_0^F}^{G^F}(\tilde{\mu})) = 0$ if λ,μ are not $W(T_0)^F$-conjugate.

<u>Proof</u>. This follows from Theorem 3.1, using the Bruhat decomposition $G^F = \bigcup_{w \in W(T_0)^F} B_0^F \dot{w} B_0^F$ (see Proposition 2.12).

<u>Example</u>. $G = GL_2$, $G^F = GL(2,q)$. Here we have $T_0^F = \{ \begin{pmatrix} \gamma^a & 0 \\ 0 & \gamma^b \end{pmatrix} \}$ where γ is a generator of $\mathbb{F}_q^*$. So the characters of T_0^F are given by $\theta_{m,n} : \begin{pmatrix} \gamma^a & 0 \\ 0 & \gamma^b \end{pmatrix} \to \gamma^{ma} \cdot \gamma^{nb}$, where we also denote a complex primitive $(q-1)$ st root of unity by γ. We have $B_0^F = \{ \begin{pmatrix} \gamma^a & * \\ 0 & \gamma^b \end{pmatrix} \}$ and we give below the values of $\mathrm{Ind}_{B_0^F}^{G^F}(\tilde{\theta}_{m,n}) = \phi_{m,n}$, when $\phi_{m,n}$ is irreducible, at representatives of the conjugacy classes of G^F. Here the element in the last column is conjugate to $\begin{pmatrix} \eta^a & 0 \\ 0 & \eta^{aq} \end{pmatrix}$ over $\mathbb{F}_{q^2}$, where η is a generator of $\mathbb{F}_{q^2}^*$. In the table η also denotes

a complex primitive (q^2-1) st root of unity. The second row denotes a family of irreducible characters of G^F known as the discrete series. This family of characters cannot be constructed in a straightforward way like the principal series, and corresponds in some sense to the other "non-split" torus of G^F. A precise definition of the discrete series will be given in the next chapter.

Class representative	$\begin{pmatrix}1&0\\0&1\end{pmatrix}$	$\begin{pmatrix}\gamma^a&0\\0&\gamma^a\end{pmatrix}$	$\begin{pmatrix}\gamma^a&1\\0&\gamma^a\end{pmatrix}$	$\begin{pmatrix}\gamma^a&0\\0&\gamma^b\end{pmatrix}$	$\begin{pmatrix}\eta^a&0\\0&\eta^{aq}\end{pmatrix}$
		$(\gamma^a\neq 1)$		$(\gamma^a\neq\gamma^b)$	$(a\not\equiv 0(q+1))$
Number of classes	1	q-2	q-1	$\frac{1}{2}(q-1)(q-2)$	$\frac{1}{2}q(q-1)$
$\phi_{m,n}$ $(m,n=1,2,\ldots,q-1,\ m\neq n)$	q+1	$(q+1)\gamma^{(m+n)a}$	$\gamma^{(m+n)a}$	$\gamma^{ma+nb}+\gamma^{na+mb}$	0
ψ_m $(m=1,2,\ldots,q^2-2$ m not a multiple of $q+1)$	q-1	$(q-1)\gamma^{ma}$	$-\gamma^{ma}$	0	$-(\eta^{ma}+\eta^{maq})$

Some remarks on inducing from Borel subgroups.

1. The construction of inducing from a Borel subgroup (or more

generally from a parabolic subgroup) occurs also in the case of real semisimple Lie groups (from where the term "principal series" arises) and in the case of semisimple p-adic groups. (See e.g. [43], pp. 16, 123.)

2. Consider the reductive algebraic group G over $K = \bar{\mathbb{F}}_q$. Let $\lambda \in X(T_0)$. Let V be the space of all morphisms $f : G \to K$ such that $f(gb) = \tilde{\lambda}(b^{-1})f(g)$ $(g \in G, b \in B_0)$, where λ is extended to a character $\tilde{\lambda}$ of B_0 (into K^*) as usual. G acts on V by left translations and this gives the representation $\mathrm{Ind}_{B_0}^{G}(\tilde{\lambda})$. These representations play a big role in the "modular" representation theory of G (ie. rational representations over K). All irreducible modular representations of G are realized inside spaces of the form V. (See [80], p. 85 or [81], p. 210.)

3. Let $\mathfrak{g}$ be a complex semisimple Lie algebra, $\mathfrak{h}$ a Cartan subalgebra, $\mathfrak{b}$ a Borel subalgebra with $\mathfrak{b} = \mathfrak{h} \oplus n$ where n is nilpotent. Let $U(\mathfrak{g})$ be the universal enveloping algebra of $\mathfrak{g}$. If λ is a linear functional on $\mathfrak{h}$, we can lift it to a linear functional on $\mathfrak{b}$ by setting $\lambda(h+n) = \lambda(h)$, if $h \in \mathfrak{h}$ and $n \in n$. If L is a $\mathfrak{b}$-module for λ the induced $\mathfrak{g}$-module is defined as $U(\mathfrak{g}) \otimes_{U(\mathfrak{b})} L$ (regarding $U(\mathfrak{g})$ as a right $U(\mathfrak{b})$-module). Now if we take L to be a $\mathfrak{b}$-module for $\lambda-\delta$ (where δ is half the sum of the positive roots) the induced module is called the Verma module of $\mathfrak{g}$ associated with λ and denoted by $M(\lambda)$. These objects have been studied a great deal in recent years. (See [20], Chapter 7.)

CHAPTER IV. DISCRETE SERIES REPRESENTATIONS AND HARISH-CHANDRA THEORY.

We recall (Chapter I) that if P is a parabolic subgroup of G then we have a Levi decomposition $P = LV$. If P is F-stable then so is V and we chan choose L to be F-stable by choosing it to contain an F-stable maximal torus T. Then we have $P^F = L^F V^F$ and $V^F = O_p(P^F)$. By a parabolic subgroup of G^F we mean a subgroup of the form P^F; we will refer to V^F as its unipotent radical and to L^F as a Levi subgroup. We will also denote the lift of a representation ρ of L^F to P^F by $\tilde{\rho}$. The representations of L^F which are best suited for lifting to P^F and inducing to G^F are those which "do not come from below", ie. the so-called cuspidal or discrete series representations.

The results to be described here are due to Harish-Chandra, and can be found in [12], , [7], or [72].

<u>Definition</u>. Let ζ be an irreducible representation (or character) of G^F. We say ζ is cuspidal or is in the discrete series, if for any $P^F \neq G^F$, $\zeta|V^F$ does not contain the trivial representation (or character) of V^F.

<u>Remark</u>. If G is a torus, we define every irreducible representation of G^F to be cuspidal.

<u>Notation</u>. $\mathcal{E}(G)$ (resp. ${}^{\circ}\mathcal{E}(G)$) is the set of isomorphism classes of irreducible representations (resp. cuspidal representations) of G^F.

<u>Lemma 4.1</u>. Let ζ be in $\mathcal{E}(G)$. Then ζ is cuspidal if and only if the following holds.

(4.2) For every $x \in G^F$ we have $\sum_{v \in V^F} \zeta(xv) = 0$ for all

$P^F \neq G^F$.

Proof. Let $e = \frac{1}{|V^F|} \sum_{v \in V^F} v$. Then e is an idempotent in $\mathbb{C}[V^F]$ affording the trivial representation. So if (4.2) is satisfied then $\zeta(e) = 0$ and ζ is cuspidal. Conversely, let ζ be cuspidal. Then $\zeta(e) = 0$, hence $\zeta(xe) = 0$ for all $x \in G^F$ and (4.2) holds.

Remark. More generally a class function f on G^F is said to be cuspidal if (4.2) holds. In this case it is not enough to check that $f(e) = 0$ to show that f is cuspidal.

Proposition 4.3. Let $\zeta \in \mathcal{E}(G)$. Then there is a parabolic subgroup P^F of G^F and $\psi \in {}^\circ\mathcal{E}(L)$ such that ζ is a constituent of $\mathrm{Ind}_{P^F}^{G^F}(\tilde{\psi})$.

Proof. There exist parabolics P such that ζ occurs in $\mathrm{Ind}_{V^F}^{G^F}(1)$; for example, we could take $P = G$, in which case $V = 1$. Take a minimal such P, and consider $\zeta | P^F$. Let M be a G^F-module affording ζ and let N be the P^F-submodule of M consisting of the fixed points under V^F. By Frobenius reciprocity $N \neq 0$. Consider the representation $\mathcal{G}$ of P^F on N. This is the direct sum of irreducible representations which must all be lifts of cuspidal representations of L^F; otherwise, there would be a proper parabolic subgroup of L^F whose unipotent radical has a fixed point on N, and this would lead to a parabolic subgroup of G^F which is properly contained in P^F, and such that ζ restricted to its unipotent radical contains 1, contradicting the minimality of P. Since ζ occurs in $\mathrm{Ind}_{V^F}^{G^F}(1)$ it follows now that it occurs in some $\mathrm{Ind}_{P^F}^{G^F}(\tilde{\tau})$ where τ is a cuspidal representation of L^F.

Remark. In 4.3 we could have $P^F = G^F$, in which case $\zeta \in {}^{\circ}\mathcal{E}(G)$. If P is an F-stable parabolic subgroup of G, let $\mathcal{E}(G,P)$ be the set of all ϕ in $\mathcal{E}(G)$ which occur as constituents of $\mathrm{Ind}_{P^F}^{G^F}(\tilde{\psi})$ for some ψ in ${}^{\circ}\mathcal{E}(L)$. Now it follows that if we choose representatives $\{P_i\}$ for the conjugacy classes of F-stable parabolic subgroups of G then $\mathcal{E}(G) = \bigcup_i \mathcal{E}(G,P_i)$. We would like to refine this decomposition and get a disjoint union.

Definition. Two F-stable parabolic subgroups P and P' are said to be associated if there exist F-stable Levi subgroups L,L' of P,P' and $x \in G^F$ such that ${}^{x}L^F = L'^F$.

Our aim is to show that $(\mathrm{Ind}_{P^F}^{G^F}(\tilde{\zeta}), \mathrm{Ind}_{P'^F}^{G^F}(\tilde{\zeta}'))_{G^F} = 0$ unless P and P' are associated, where ζ,ζ' are in ${}^{\circ}\mathcal{E}(L)$, ${}^{\circ}\mathcal{E}(L')$ respectively.

Example. Let $G = GL_n$. Then two F-stable parabolic subgroups P and P' are associated if and only if their Levi subgroups are both isomorphic to $GL_{n_1} \times GL_{n_2} \times \ldots \times GL_{n_k}$ for the same set of positive integers $n_1, n_2, \ldots, n_k$. The next lemma is rather technical and we will omit the proof. A proof can be found in [12], where the finite group G^F is considered throughout without any reference to the algebraic group (the BN-pair axioms serving as a substitute). This lemma is proved there for two parabolics P_1^F, P_2^F which contain a common Borel subgroup B^F. Since any two F-stable Borel subgroup are conjugate by an element of G^F (see 2.8) there is no loss of generality in assuming that the two parabolic subgroups of G that we are considering do in fact contain a common F-stable Borel subgroup.

Lemma. 4.4. Let P_1, P_2 be F-stable parabolic subgroups containing B_0 with $P_i = L_iV_i$ $(i=1,2)$, the L_i being F-stable. Then we have the following.

(i) $P_1^F \cap P_2^F = (L_1^F \cap L_2^F)(L_1^F \cap V_2^F)(L_2^F \cap V_1^F)(V_1^F \cap V_2^F)$.
Furthermore, we have uniqueness of expression for elements of $P_1^F \cap P_2^F$.

(ii) The unipotent radical of $L_1^F \cap P_2^F$ is $L_1^F \cap V_2^F$ and the unipotent radical of $L_2^F \cap P_1^F$ is $L_2^F \cap V_1^F$.

(iii) $P_1^F \cap V_2^F \subset V_1^F$ implies $L_1^F \subset L_2^F$.

Theorem 4.5. Let P_1, P_2 be as in Lemma 4.4 and let $W = W(T_0)$. Let $\psi_i \in {}^\circ\mathcal{E}(L_i)$ $(i=1,2)$. Then $(\mathrm{Ind}_{P_1^F}^{G^F}(\tilde{\psi}), \mathrm{Ind}_{P_2^F}^{G^F}(\tilde{\psi}))_{G^F} = 0$

unless $L_1^F = {}^{\dot w}L_2^F$ and $\psi_1 = {}^{\dot w}\psi_2$ for some $w \in W^F$ (in particular, unless P_1, P_2 are associated).

Proof. By the Bruhat decomposition (2.11) and Mackey's Theorem (3.1) we have

$$(\mathrm{Ind}_{P_1^F}^{G^F}(\tilde{\psi}_1), \mathrm{Ind}_{P_2^F}^{G^F}(\tilde{\psi}_2))_{G^F} = \sum_w (\tilde{\psi}_1, {}^{\dot w}\tilde{\psi}_2)_{P_1^F \cap {}^{\dot w}P_2^F},$$

where w runs over a subset of W^F. Now consider

$$(\tilde{\psi}_1, {}^{\dot w}\tilde{\psi}_2) = \frac{1}{|P_1^F \cap {}^{\dot w}P_2^F|} \sum_{x,y,z,v} \tilde{\psi}_1(xyzv)\, \overline{{}^{\dot w}\tilde{\psi}_2(xyzv)},$$

the sum being over all $x \in L_1^F \cap L_2^F$, $y \in L_1^F \cap {}^{\dot w}V_2^F$, $z \in V_1^F \cap {}^{\dot w}L_2^F$, $v \in V_1^F \cap {}^{\dot w}V_2^F$ (using 4.4 (i)). Now since $\tilde{\psi}_1$ and ${}^{\dot w}\tilde{\psi}_2$ are trivial on $V_1^F \cap {}^{\dot w}V_2^F$, The above expression is a multiple of $\sum_{x,y,z} \tilde{\psi}_1(xyz)\, \overline{{}^{\dot w}\tilde{\psi}_2(xyz)}$, and hence of $\sum_{x,y,z} \tilde{\psi}_1(xy)\, {}^{\dot w}\tilde{\psi}_2(xz)$ (writing $xyz = xz(z^{-1}yz)$ and noting that $z^{-1}yz \in {}^{\dot w}V_2^F$, $z \in V_1^F$). Now using (4.4) (ii), (4.2) and the fact that $\psi_1 \in {}^{\circ}\mathcal{E}(L_1)$ we see that $\sum_y \tilde{\psi}_1(xy) = 0$ unless $L_1^F \cap {}^{\dot w}V_2^F = 1$. Similarly $\sum_z {}^{\dot w}\psi_2(xz) = 0$ unless $V_1^F \cap {}^{\dot w}L_2^F = 1$. Hence we have

$$(\mathrm{Ind}_{P_1^F}^{G^F}(\tilde{\psi}_1), \mathrm{Ind}_{P_2^F}^{G^F}(\tilde{\psi}_2)) = 0 \text{ unless for some } w \in W^F,$$

$L_1^F \cap {}^{\dot w}V_2^F$ and $V_1^F \cap {}^{\dot w}L_2^F$ are both trivial. But then $P_1^F \cap {}^{\dot w}V_2^F \subset V_1^F$ and ${}^{\dot w^{-1}}V_1^F \cap P_2^F \subset V_2^F$. By (4.4) (iii) this implies that $L_1^F \subset {}^{\dot{w}}L_2^F$ and $L_2^F \subset {}^{\dot{w}^{-1}}L_1^F$ and thus $L_1^F = {}^{\dot{w}}L_2^F$, as required. Also it now follows that $\psi_1 = {}^{\dot{w}}\psi_2$, as otherwise $(\tilde{\psi}_1, {}^{\dot{w}}\tilde{\psi}_2)_{P_1^F \cap {}^{\dot{w}}P_2^F} = 0$.

<u>Remark</u>. In fact 4.4(iii) can be sharpened to: if $P_1^F \cap V_2^F \subset V_1^F$, then $L_1 \subset L_2$ (see [72], p. 627). So in Theorem 4.5 we can prove the scalar product is zero unless $L_1 = {}^{\dot{w}}L_2$.

Now we have shown that

(4.6) $$\mathcal{E}(G) = \bigcup_P \mathcal{E}(G,P),$$

where the union is disjoint and P runs over a set of representatives for the associativity classes of F-stable parabolic subgroups of G. Thus the problem of determining $\mathcal{E}(G)$ is reduced to two problems. (1) Find ${}^{o}\mathcal{E}(G)$. (2) Decompose $\mathrm{Ind}_{P^F}^{G^F}(\tilde{\psi})$ where $\psi \in {}^{o}\mathcal{E}(L)$. These two problems will be discussed in subsequent chapters. Neither of them is completely solved, but a large number of the representations in ${}^{o}\mathcal{E}(G)$ have been constructed by Lusztig and Deligne [18] and by Lusztig in [45]. The centralizer algebra of $\mathrm{Ind}_{P^F}^{G^F}(\tilde{\psi})$ has been studied in some cases and this will be discussed in Chapter VIII.

CHAPTER V. THE ℓ-ADIC COHOMOLOGY

In this chapter we will first give a review of the definitions and elementary properties of sheaves on a topological space. This material is introduced so that one knows what to expect in the case of ℓ-adic sheaves. Then we define schemes and the sheaf cohomology of schemes. After this we describe ℓ-adic sheaves and ℓ-adic cohomology and give a survey of the main properties of ℓ-adic cohomology groups of a scheme which is separated and of finite type over an algebraically closed field. We point out the analogous results in the classical case and give references to the results in both the ℓ-adic and the classical cases.

§1. Sheaves (Classical theory). (See [29], [32] or [51].)

Let X be a topological space. A presheaf Φ of abelian groups (or sets, rings, etc.) consists of the following data. For every open subset U of X we have an abelian group (or set, ring, etc.) $\Phi(U)$, together with (restriction) homomorphisms $\rho_{UV}: \Phi(U) \to \Phi(V)$ whenever $V \subset U$, subject to the conditions (1) $\Phi(\emptyset) = 0$, (2) ρ_{UU} is the identity map of U, and (3) if $U \supset V \supset W$, then $\rho_{UW} = \rho_{VW}\rho_{UV}$.

The presheaf Φ is said to be a sheaf if it also satisfies:

(5.1) For each open set U in X and each open covering $\{U_\alpha\}$ of U, and each family $\{s_\alpha\}$ such that $s_\alpha \in \Phi(U_\alpha)$ and s_α, s_β have the same restriction to $\Phi(U_\alpha \cap U_\beta)$ for all α, β, there is a unique $s \in \Phi(U)$ whose restriction to U_α is s_α for all α.

If $x \in X$, the <u>stalk</u> of Φ at x is defined to be $\varinjlim_{\substack{x \in U \\ U \text{ open}}} \Phi(U)$, and is denoted by Φ_x.

We define a morphism $\emptyset\colon \Phi \to g$ of presheaves or sheaves on X as a family of homomorphisms $\emptyset(U)\colon \Phi(U) \to g(U)$ (U open in X) commuting with the restriction homomorphisms. Then $\emptyset$ induces a homomorphism $\Phi_x \to g_x$ of the stalks at any point x of X.

Given a presheaf Φ on X, there is a canonical sheaf $\tilde{\Phi}$, called its sheafification, and a morphism $\Phi \to \tilde{\Phi}$ (of presheaves) through which all morphisms from Φ into sheaves factor uniquely. We can construct this as follows ([51], p. 30). Let E be the disjoint union of the stalks Φ_x for all $x \in X$. Given $s \in \Phi(U)$, define $\tilde{s}\colon U \to E$ by $\tilde{s}(x) = s_x$, where, in the canonical map $\Phi(U) \to \Phi_x$, s maps on s_x. Thus $\tilde{s}$ is a section of E over U. Give E the coarsest topology such that all the maps $\tilde{s}$ are continuous. Define, for each open U, $\tilde{\Phi}(U)$ to be the set of continuous sections of E over U. Then $\tilde{\Phi}$ is the required sheaf. If we had started with a presheaf Φ which was already a sheaf, $\tilde{\Phi}$ would be isomorphic to Φ. Because of this way of viewing a sheaf on X, the elements of $\Phi(U)$ for any sheaf Φ are called the sections of Φ over U. The elements of $\Phi(X)$ are called global sections, and often $\Phi(X)$ is denoted by $\Gamma(X,\Phi)$.

Another way of thinking of a presheaf is as a contravariant functor from the category $C(X)$ whose objects are the open sets in X and morphisms are inclusions of open sets, into the category of abelian groups (or sets, rings, etc.). This is the point of view which will be useful later.

Direct and inverse images of sheaves ([32], p. 65).

Let $f: X \to Y$ be a continuous map and Φ a sheaf on X. Then the direct image $f_*\Phi$ is the sheaf on Y defined by $(f_*\Phi)(V) = \Phi(f^{-1}(V))$ for any open set V of Y. If $\mathcal{G}$ is a sheaf on Y, the inverse image $f^*\mathcal{G}$ is the sheaf on X associated with the presheaf p where $p(U) = \varinjlim_{\substack{V \supset f(U) \\ V \text{ open}}} \mathcal{G}(V)$, for an open set U in X. We then have $\mathrm{Hom}_Y(\mathcal{G}, f_*\Phi) = \mathrm{Hom}_X(f^*\mathcal{G}, \Phi)$, i.e. the functor f^* from the category of sheaves of abelian groups on Y to the category of sheaves of abelian groups on X is the left adjoint of the functor f_* from the category of sheaves of abelian groups on X to the category of sheaves of abelian groups on Y.

We also note that subsheaves and quotient sheaves can be defined for sheaves of abelian groups on X ([32], p. 64).

Restriction ([32], p. 65).

Let Z be a subspace of X with the induced topology and let $i: Z \to X$ be the inclusion map. If Φ is a sheaf on X, the sheaf $i^*\Phi$ on Z is called the restriction of Φ to Z and denoted by $\Phi|Z$. We note that the stalk of $\Phi|Z$ at any $x \in Z$ is just Φ_x.

Extension by zero ([32], p. 68).

Let $j: U \to X$ be an open immersion (i.e., j is an isomorphism of U with an open subset of X). If Φ is a sheaf on

U, the sheaf on X associated with the presheaf which attaches $\Phi(V)$ to V if $V \subset U$ or 0 to V if $V \not\subset U$ is called the extension of Φ by zero and is denoted by $j!\Phi$. If $Z = X - U$ and $i: Z \to X$ is inclusion, we have an exact sequence $0 \to j!(\Phi|U) \to \Phi \to i_*(\Phi|Z) \to 0$ of sheaves on X for any sheaf Φ on X.

Examples.

1. Constant sheaf. Let A be an abelian group, given the discrete topology. The constant sheaf $\mathcal{A}$ on X is defined by $\mathcal{A}(U)$ = {Continuous maps of U into A}. So if U is connected $\mathcal{A}(U) = A$. In fact, $\mathcal{A}$ is the sheafification of the presheaf which assigns A to every open set U of X.

2. Locally constant sheaf. Φ is a locally constant sheaf on X if there is a covering of X by open sets $\{U_i\}$ such that $\Phi|U_i$ is a constant sheaf for each i.

3. Let X be an affine variety over the algebraically closed field K, as in Chapter I. We say a function $f: X \to K$ is regular at $a \in X$ if there is a neighborhood V of a in which $f = g|h$ where $g,h \in K[X]$ and $h \neq 0$ in V. We say f is regular if it is regular at all $a \in X$. If U is open in X, let $\Phi(U)$ be the set of all regular functions on U. Then Φ is a sheaf of rings on X. Similarly we have a naturally defined sheaf on any quasiprojective variety X ([32], p. 62).

Sheaf cohomology ([32, pp. 202-208).

The category of sheaves of abelian groups on a topological space X is an abelian category which has enough injectives;

hence every sheaf Φ on X has an injective resolution. Furthermore, the global section functor $\Gamma: \Phi \to \Gamma(X,\Phi) = \Phi(X)$ is left exact from this category into the category of abelian groups. Thus we can take the derived functors of Γ and define the cohomology groups of X with coefficients in Φ. In other words, if

$$o \to \Phi \to C^0 \to C^1 \to \cdots \to C^i \to C^{i+1} \quad \cdots$$

is an injective resolution of Φ, we consider the complex

$$o \to \Gamma(X,\Phi) \to \Gamma(X,C^0) \to \Gamma(X,C^1) \to \cdots \to \Gamma(X,C^i) \xrightarrow{d_i} \Gamma(X,C^{i+1}) \to \cdots$$

and define $H^i(X,\Phi) = \ker d_i/\mathrm{Im}\, d_{i-1}$. (Note that $H^0(X,\Phi) = \Gamma(X,\Phi)$.)

The sequence of functors $R^i\Gamma: \Phi \to H^i(X,\Phi)$ from the category of sheaves of abelian groups on X into the category of abelian groups forms a "δ-functor" in the sense that given any short exact sequence $o \to \Phi' \to \Phi \to \Phi'' \to o$ of sheaves we have a long exact sequence

$$o \to H^0(X,\Phi') \to H^0(X,\Phi) \to H^0(X,\Phi'') \xrightarrow{\delta_1} H^1(X,\Phi') \to H^1(X,\Phi)$$
$$\to H^1(X,\Phi'') \xrightarrow{\delta_2} \cdots$$

Furthermore, it is a "universal δ-functor" in the sense that given any other sequence of functors (T_i) with the above property, and a morphism of functors $f^0: \Gamma \to T_0$, there is a unique sequence of morphisms $f^i: R\Gamma^i \to T_i$ for each i, starting with the given f_0, which commute with the δ_i for each short exact sequence.

<u>Higher direct images of a sheaf</u> ([32], p. 250).

Let $f: X \to Y$ be a continuous map and Φ a sheaf on X.

Then the ith direct image sheaf $R^i f_* \Phi$ on Y is the sheafification of the presheaf on Y which assigns to any open set V in Y the abelian group $H^i(f^{-1}(V), \Phi|f^{-1}(V))$.

§2. <u>Schemes</u> ([51], Ch. 6; [32], Chapter II; [54], Ch. 2).

In this section we define and give a brief review of schemes, without too many technical details.

A <u>ringed space</u> is a pair (X, O_X) consisting of a topological space X and a sheaf of rings O_X (called the structure sheaf) on X. A morphism $(X, O_X) \to (Y, O_Y)$ consists of a continuous map from X to Y and a morphism of sheaves $O_Y \to O_X$. Thus for each open set V in Y we have a ring homomorphism $O_Y(V) \to O_X(f^{-1}(V))$, compatible with restrictions, and hence an induced homomorphism of stalks $O_{Y,f(x)} \to O_{X,x}$ for any $x \in X$. A <u>locally ringed space</u> is a ringed space (X, O_X) in which for each $x \in X$ the stalk $O_{X,x}$ is a local ring, and a morphism of locally ringed spaces is one in which the homomorphism of stalks mentioned above is a local homomorphism of local rings, i.e., the inverse of the maximal ideal of $O_{X,x}$ is the maximal ideal of $O_{Y,f(x)}$. (See [51], p. 34, or [32], p. 72.)

Let A be a commutative ring with 1. Let $X = \operatorname{Spec} A$ be the set of all prime ideals of A. For any $E \subset A$ we let $V(E) \subset X$ be the set of all prime ideals containing E. We define a topology (the Zariski topology) on X by letting the sets $V(E)$ be the closed sets. One can define a sheaf of rings O_X on X in such a way that the stalk $O_{X,x}$ of O_X

at $x \in X$ is the local ring of A corresponding to the prime ideal x of A. Then Spec A is a locally ringed space. (See [51], p. 36; [32], p. 70.)

An *affine scheme* is a locally ringed space which is isomorphic to Spec A for some A. Finally a *scheme* is a locally ringed space (X,O_X) in which every point has an open neighborhood U such that the pair $(U,O_X|U)$ is an affine scheme. A morphism of schemes is a morphism regarding them as locally ringed spaces. (See [51], p. 43 or [52], p. 74, where there are also several examples of schemes.)

If S is a fixed scheme, a *scheme over* S is a scheme X together with a morphism $X \to S$. Often one considers the category of schemes over a fixed base scheme S, the morphisms from X to Y being given by commutative triangles

$$\begin{array}{ccc} X & \rightarrow & Y. \\ & \searrow \quad \swarrow & \\ & S & \end{array}$$

If $S = \text{Spec } A$, we talk of a scheme over A. Any scheme X can be regarded as a scheme over $\mathbb{Z}$; it is enough to show this for affine schemes, and given any ring A with 1 the natural map $\mathbb{Z} \to A$ gives rise to a morphism Spec $A \to$ Spec $\mathbb{Z}$.

In the category of schemes over S, products exist. In other words, given $X \to S$, $Y \to S$ we have a product scheme $X \underset{S}{\times} Y$ over S and a commutative diagram

$$\begin{array}{ccc} & X \underset{S}{\times} Y & \\ {}^{p_1}\swarrow & & \searrow^{p_2} \\ X & & Y \\ & \searrow \quad \swarrow & \\ & S & \end{array}$$

such that given any scheme Z over S and morphisms $f: Z \to X$, $g: Z \to Y$ which make a commutative diagram with the given morphisms $X \to S$ and $Y \to S$, there is a unique morphism

$\theta: Z \to X \underset{S}{\times} Y$ such that $f = p_1 \cdot \theta$ and $g = p_2 \cdot \theta$. If $S = \operatorname{Spec} \mathbb{Z}$ we write $X \times Y$ for $X \underset{\mathbb{Z}}{\times} Y$. (See [32], p. 87.)

Remarks. 1. In the category of sets, if $X \to S$, $Y \to S$ are inclusions, then $X \underset{S}{\times} Y$ is just $X \cap Y$. Thus when we introduce "generalized topologies" (e.g., the étale topology) on a scheme X, the product $X \times Y$ will take the place of $X \cap Y$.

2. If X,Y are schemes over S, the map $X \underset{S}{\times} Y \to Y$ makes $X \underset{S}{\times} Y$ a scheme over Y, and we say it is obtained from X by base extension.

Example. Let X_0 be an irreducible affine variety over the algebraically closed field K, as in Chapter I. Then X_0 can be enlarged to an affine scheme X over K; in fact, $X = \operatorname{Spec} A$ where $A = K[X_0]$. The idea is that X consists of not only the points of X_0 (which correspond to maximal ideals of A and represent the closed points of X) but also of points corresponding to each irreducible subvariety of X_0. The morphism $X \to \operatorname{Spec} K$ consists of mapping X onto Spec K (which consists of a single point) and K into A in the natural way. Similarly we can associate a scheme with any quasi-projective variety. (See [32], p. 78.) We will call such a scheme a quasiprojective scheme.

Notation. If $x \in X$, where X is a scheme, the stalk of O_X at x is a local ring whose residue field will be denoted by $k(x)$.

Fibres of a morphism

Suppose we have a morphism $f: X \to Y$ of schemes and $y \in Y$. Then the fibre of f over y is defined to be the scheme $X \times_Y \operatorname{Spec} k(y)$. (Note that there is a natural morphism $\operatorname{Spec} k(y) \to Y$, which takes $\operatorname{Spec} k(y)$ onto y and the stalk $O_{Y,y}$ onto $k(y)$.) This is a scheme over $k(y)$ whose underlying topological space is homeomorphic to $f^{-1}(y)$, the usual fibre over y ([32], p. 89, or [56], p. 16). We sometimes denote the fibre of f over y by X_y.

"Points," geometric points and geometric fibres

If X_0 is an affine variety over an algebraically closed field K, the points of X_0 are in bijection with the K-homomorphisms of $A = K[X_0]$ into K. If X,Y are schemes over S we define a Y-valued point of X to be a morphism (over S) from Y into X. In particular, a geometric point of a scheme X is a morphism $\operatorname{Spec} \tilde{K} \to X$ where $\tilde{K}$ is an algebraically closed field. The image of this morphism is called the center of the geometric point. If $x \in X$ then we have a morphism $\operatorname{Spec} \overline{k(x)} \to X$ which is a geometric point centered at x. Conversely, any geometric point centered at $x \in X$ is a morphism of the form $\operatorname{Spec} \tilde{K} \to X$ where $\tilde{K} \supset \overline{k(x)}$. [For a discussion of points and the motivation for defining them in this way, see [54], p. 218.]

Suppose we have a morphism $f: X \to Y$ and $\bar{x}$ is a geometric point of Y, i.e., we have a morphism $\bar{x}: \operatorname{Spec} \tilde{K} \to Y$. Then the "geometric fibre" of f over $\bar{x}$ is defined to be $X \times_Y \operatorname{Spec} \tilde{K}$, and denoted by $X_{\bar{x}}$. Suppose $\bar{x}$ is centered at $x \in Y$. Then the connection between $X_{\bar{x}}$ and the usual fibre X_x is that there is an action of a Galois group (essentially, of $\overline{k(x)}$ over $k(x)$) and X_x is essentially a quotient of $X_{\bar{x}}$

under this action.

We now define some important kinds of morphisms of schemes.

Definition. Let $f: X \to Y$ be a morphism of schemes. Let Δ be the "diagonal morphism" $X \to X \times_Y X$ whose composition with both the projection maps $X \times_Y X \to X$ is the identity map $X \to X$. We say f is separated (or X is separated over Y) if Δ is a closed immersion. (See [32], p. 96.)

This is the analogue of the Hausdorff axiom for topological spaces. For example, any morphism $X \to Y$ of affine schemes is separated.

Definition. A morphism $f: X \to Y$ of schemes is of finite type (or X is of finite type over Y) if there is a covering of Y by open affine subsets $V_i = \operatorname{Spec} B_i$ such that each $f^{-1}(V_i)$ can be covered by a finite number of open affine subsets $U_{ij} = \operatorname{Spec} A_{ij}$, such that each A_{ij} is a finitely generated B_i-algebra. If each $F^{-1}(V_i)$ is itself affine and equal to Spec A_i, where A_i is integral over B_i then we say f is finite. (See [32], p. 84; cf. the definition of a finite morphism in Chapter I.)

Definition. A morphism $f: X \to Y$ is universally closed if it is closed (i.e., the image under f of any closed subset of X is closed) and for any morphism $Y' \to Y$, the corresponding morphism $X \times_Y Y' \to Y'$ is closed.

Definition. A morphism $f: X \to Y$ is proper if it is separated, of finite type, and universally closed. We also say X is proper over Y. For example, the affine line over a field k is not proper over k, but the projective line is proper over k. More generally, any projective variety is proper over k. Thus "proper" is the analogue of compactness in the classical case. (See [32], p. 100.)

Functoriality of Sheaf Cohomology

Let X be a scheme, Φ a sheaf of abelian groups on X. Regarding Φ as a sheaf on the underlying (Zariski) topological space of X, we can form the cohomology groups $H^i(X,\Phi)$ as described earlier.

Suppose we have a morphism $f: X \to X'$ of schemes and Φ is a sheaf on X'. By the definition of the inverse image sheaf $f^*\Phi$ on X, we have a map $\Gamma(X',\Phi) \to \Gamma(X,f^*\Phi)$, i.e., a map $H^0(X',\Phi) \to H^0(X,f^*\Phi)$. Also the functor $\Phi \to H^i(X,f^*\Phi)$ is a δ-functor from the category of sheaves on X' into the category of abelian groups. Thus, since the $R^i\Gamma$ form a universal δ-functor form the category of sheaves on X' into the category of abelian groups, we have a sequence of morphisms $H^i(X',\Phi) \to H^i(X,f^*\Phi)$. In other words, the morphism $f: X \to X'$ of schemes induces a homomorphism $H^i(X',\Phi) \to H^i(X,f^*\Phi)$ for each i.

§3. Étale cohomology ([62], Expose XVII; [63]; [64] Expose VI)

The étale topology (see [53], [2], or [63]).

Let X be a set. To give a topology on X is equivalent to giving a category $C(X)$ whose objects are the open sets in X and morphisms are inclusions of open sets. Grothendieck's idea is to generalize this notion; thus a generalized topological space in his sense is given by a category whose objects are morphisms of the form $U \to X$ in some category C, where X is a fixed object of C, and morphisms are commutative triangles $\begin{smallmatrix} U & \to & V \\ & \searrow \swarrow & \\ & X & \end{smallmatrix}$. One such "topology" is the étale topology, where C is the category of schemes, which will be described below.

<u>From now on, unless otherwise stated, we will consider only schemes which are separated and of finite type over an algebraically closed field K</u>. We can think of such a scheme as obtained by gluing together a finite number of affine schemes of the form Spec A where A is a finitely generated K-algebra.

We introduce the concept of an étale morphism. Let $f: X \to Y$ be a morphism of schemes. Then we have a morphism of sheaves $O_Y \to O_X$, and, for each $x \in X$, we have an induced map of stalks $O_{Y,f(x)} \to O_{X,x}$ which is a map of local rings.

<u>Definition</u>. f is étale if for each closed point y of Y, $f^{-1}(y)$ is finite and for each $x \in f^{-1}(y)$ the morphism of stalks $O_{Y,y} \to O_{X,x}$ gives rise to an isomorphism of the completions (with respect to their maximal ideals) $\hat{O}_{Y,y} \tilde{\to} \hat{O}_{X,x}$.

Roughly, an étale morphism is analogous to a local homeomorphism for analytic spaces over $\mathbb{C}$.

Given a scheme X, the category X_{et} is the category whose objects are étale morphisms $U \to X$ where U is a scheme, and morphisms are commutative triangles $\begin{smallmatrix} U & \to & V \\ & \searrow \swarrow & \\ & X & \end{smallmatrix}$. Then "coverings" in this "étale topology" on X consist of finite sets of morphisms $U_\alpha \xrightarrow{P_\alpha} U$ such that U is the union of the $P_\alpha(U_\alpha)$.

A presheaf of abelian groups (sets, rings, etc.) on X_{et} is a contravariant functor Φ from X_{et} into the category of abelian groups (sets, rings, etc.). Then Φ is a sheaf if it satisfies a formal analogue of (5.1), where, for any U_α, U_β, the intersection $U_\alpha \cap U_\beta$ is replaced by $U_\alpha \times_U U_\beta$. In other words, for any covering $\{U_\alpha \to U\}$, the diagram

$$\Phi(U) \to \prod_\alpha \Phi(U_\alpha) \rightrightarrows \prod_{\alpha,\beta} \Phi(U_\alpha \times_U U_\beta) \quad \text{is exact.}$$

(Here a diagram $A \xrightarrow{f} B \underset{h}{\overset{g}{\rightrightarrows}} C$ of sets and mappings is said to be exact if f maps A bijectively on the set of all $x \in B$ such that $g(b) = h(b)$.) The maps in the diagram are obtained from using the functor Φ on the morphisms $U_\alpha \to U$, $U_\alpha \times_U U_\beta \to U_\alpha$, $U_\alpha \times_U U_\beta \to U_\beta$. By abuse of language we will from now on talk of a sheaf on X.

Now we consider the global section functor $\Gamma\colon \Phi \to \Phi(X)$ from the category of sheaves of abelian groups on X into the category of abelian groups and denote its ith derived functor by $R^i\Gamma$ or $H^i(X, \ )$. Then $H^i(X,\Phi)$ is the <u>étale cohomology group</u> of X with coefficients in the sheaf Φ.

<u>Cohomology with compact support</u> ([63], p. 47).

By a theorem of Nagata, there exists a "compactification" $\tilde{X}$ of the scheme X; i.e., a scheme $\tilde{X}$ proper over K and an open immersion $j\colon X \to \tilde{X}$. (For example, such a situation arises when X is a quasiprojective variety embedded in a projective variety $\tilde{X}$.)

Let Φ be a sheaf of torsion abelian groups on X. As in the classical case, we can define a sheaf $j!\Phi$ on $\tilde{X}$, the extension of Φ by 0. This is defined as follows: $\mathrm{Hom}_{\tilde{X}}(j_!\Phi, \mathcal{J}) = \mathrm{Hom}_X(\Phi, j^*\mathcal{J})$, where $\mathcal{J}$ is any sheaf on $\tilde{X}$.

Definition. $H^i_c(X, \Phi) = H^i(\tilde{X}, j_!\Phi)$.

The $H^i_c(X, \Phi)$ are the étale cohomology groups with compact support of X with coefficients in Φ. It is a deep theorem that these groups are independent of the choice of $\tilde{X}$.

Direct and inverse images of sheaves (see [63], pp. 22, 49).

Suppose we have a geometric point $\bar{x}$: Spec $\tilde{K} \to X$. Then an étale neighborhood of $\bar{x}$ consists of a commutative diagram

$$\begin{array}{ccc} \operatorname{Spec} \tilde{K} & \to & U \\ & {}_{\bar{x}}\searrow & \swarrow \\ & & X \end{array}$$

where $U \to X$ is an étale morphism. If Φ is a sheaf on X, the stalk $\Phi_{\bar{x}}$ of Φ at the geometric point $\bar{x}$ is defined to be $\varinjlim_{U} \Phi(U)$, the limit being over all étale neighborhoods of $\bar{x}$.

Let $f: X \to Y$ be a morphism of schemes and Φ a sheaf on X. The the direct image sheaf $f_*\Phi$ on Y is defined by $(f_*\Phi)(V) = \Phi(X \times_Y V)$ for every étale morphism $V \to Y$. Thus f_* is a left exact functor from the category of sheaves of abelian groups on X into the category of sheaves of abelian groups on Y. The ith derived functor of f_* is denoted by $R^i f_*$. If $\bar{y}$ is a geometric point of Y then the stalk $(R^i f_*\Phi)_{\bar{y}}$ is isomorphic to $\varinjlim_{V} H^i(X \times_Y V, \Phi)$, the limit being over all étale neighborhoods V of $\bar{y}$.

By analogy with the classical case we then define the inverse image functor f^* from the category of sheaves of abelian groups on Y into the category of sheaves of abelian groups on X as the left adjoint of f_*. Then if $\bar{x}$ is a geometric point of X and Φ is a sheaf on Y the stalk $(f^*\Phi)_{\bar{x}}$ is isomorphic to the stalk $\Phi_{f(\bar{x})}$, where $f(\bar{x})$ is the geometric point $\operatorname{Spec} \tilde{K} \xrightarrow{\bar{x}} X \xrightarrow{f} Y$.

We now define direct images of sheaves in the "compact support" case as follows. Given $f: X \to Y$ it can be shown that we have a commutative diagram

$$\begin{array}{ccc} X & \xrightarrow{f} & Y \\ & {\scriptstyle j}\searrow \;\; \nearrow{\scriptstyle \tilde{f}} & \\ & \tilde{X} & \end{array}$$

, where $\tilde{f}$ is proper. Then we define $(R^i f!)(\Phi) = R^i(\tilde{f}_*)(j!\Phi)$, where Φ is a torsion sheaf on X. (Note the notation $f!$, rather than f_*, in the compact support case.) Again it can be shown that this definition is independent of the choice of $\tilde{X}$.

<u>ℓ-adic sheaves</u> ([63], p. 82).

For any prime ℓ, let Z_ℓ, Q_ℓ denote the ring of ℓ-adic integers and the field of ℓ-adic numbers, respectively.

An <u>étale covering</u> of a scheme X is a scheme Y, and a finite étale morphism $f: Y \to X$.

<u>Definition</u>. A sheaf Φ on X is said to be <u>locally constant</u> if there is an étale covering $f: Y \to X$ such that $f^*\Phi$ is a constant sheaf on Y.

<u>Definition</u>. A sheaf Φ on X is said to be <u>constructible</u> if X is the union of a finite number of locally closed sets on each of which Φ is locally constant. (Recall that a subset of a topological space is locally closed if it is the intersection

of an open set and a closed set.)

We now define an ℓ-adic sheaf on X, and note that it is <u>not</u> a sheaf on X in the sense that we have defined earlier.

<u>Definition</u>. A Z_ℓ-sheaf (or ℓ-adic sheaf) Φ on X is a projective system of sheaves Φ_n, when Φ_n is a constructible sheaf of $\mathbb{Z}/\ell^{n+1}\mathbb{Z}$-modules such that the morphisms $\Phi_n \to \Phi_{n-1}$ factor through an isomorphism as in the commutative diagram below.

$$\begin{array}{ccc} \Phi_n \underset{\mathbb{Z}/\ell^{n+1}\mathbb{Z}}{\otimes} \mathbb{Z}/\ell^n\mathbb{Z} & \to & \Phi_n \\ & \searrow^{\wr} \quad \swarrow & \\ & \Phi_{n-1} & \end{array}$$

The stalk $\Phi_{\bar{x}}$ of a Z_ℓ-sheaf Φ on X at a geometric point $\bar{x}$ of X is defined to be the Z_ℓ-module $\varprojlim (\Phi_n)_{\bar{x}}$.

<u>Examples of Z_ℓ-sheaves</u>

(1) Φ_n is the constant sheaf $\mathbb{Z}/\ell^{n+1}\mathbb{Z}$. Then Φ is the constant Z_ℓ-sheaf Z_ℓ.

(2) Φ_n is the sheaf $\mu_{\ell^{n+1}}$ of ℓ^{n+1}th roots of unity (see below).

<u>Definition</u>. $H^i(X,\Phi) = \varprojlim H^i(X,\Phi_n)$, where Φ is a Z_ℓ-sheaf.

$$H^i_c(X,\Phi) = \varprojlim H^i_c(X,\Phi_n).$$

<u>Remark</u>. The groups $H^i(X,\Phi_n)$, $H^i_c(X,\Phi_n)$ are finite, and so there is a good definition of $\varprojlim$.

In particular, we have by definition $(H^i_c(X,Z_\ell) = \varprojlim H^i_c(X,\mathbb{Z}/\ell^n\mathbb{Z})$.

<u>Q_ℓ-sheaves.</u> The category of Q_ℓ-sheaves on X is defined to be the quotient of the category of Z_ℓ-sheaves on X by the Serre subcategory of torsion Z_ℓ-sheaves (see e.g., [24], Chapter 15).

Given a Z_ℓ-sheaf Φ we denote the corresponding Q_ℓ-sheaf by $\Phi \otimes Q_\ell$, and its stalk at the geometric point $\bar{x}$ of X is defined to be $\Phi_{\bar{x}} \otimes_{Z_\ell} Q_\ell$. Suppose Φ is a Q_ℓ-sheaf on X which is represented by a Z_ℓ-sheaf Φ', i.e., $\Phi = \Phi' \otimes Q_\ell$. Then we define

$$H_c^i(X,\Phi) = (\varprojlim H_c^i(X,\Phi' \otimes Z/\ell^n Z)) \otimes_{Z_\ell} Q_\ell.$$

In particular, $H_c^i(X,Q_\ell) = H_c^i(X,Z_\ell) \otimes_{Z_\ell} Q_\ell$.

We will be mostly concerned with the cohomology groups $H_c^i(X,Q_\ell)$, or with the (finite-dimensional) $\bar{Q}_\ell$-spaces $H_c^i(X,\bar{Q}_\ell) = H_c^i(X,Q_\ell) \otimes_{Q_\ell} \bar{Q}_\ell$, where $\bar{Q}_\ell$ is an algebraic closure of Q_ℓ.

<u>Tate twists</u> ([85], p. 96).

For every integer $n > 0$, the group of nth roots of unity in $K = \bar{\mathbb{F}}_q$ is denoted by μ_n. Let ℓ be a prime not equal to $p = \text{char } K$. Then the groups μ_{ℓ^n} form a projective system (with homomorphisms $\mu_{\ell^n} \to \mu_{\ell^{n-1}}$) and we let $H = Q_\ell \otimes_{Z_\ell} \varprojlim \mu_{\ell^n}$. The Galois group $G(K,\mathbb{F}_q)$ acts on μ_{ℓ^n} for each n, and thus it acts on the one-dimensional Q_ℓ-vector space H. For any vector space V over Q_ℓ, we define the Tate twists of V by $V(m) = V \otimes_{Q_\ell} H^{\otimes m}$, where, if $m < 0$, $H^{\otimes m}$ is defined as $\mathrm{Hom}(H^{\otimes m}, Q_\ell)$. Then if $G(K,\mathbb{F}_q)$ acts on V, it acts on $V(m)$. Similarly we define the twists of a vector space V over $\bar{Q}_\ell$.

As an example of a Q_ℓ-space V on which $G(K,\mathbb{F}_q)$ acts, consider a scheme X_0 over $\mathbb{F}_q$ and let $X = X_0 \times_{\mathbb{F}_q} \mathrm{Spec}\, K$ be the scheme over K obtained by base extension. Then $G(K,\mathbb{F}_q)$ acts on the second factor, hence on X, and hence on $V = H^i_c(X,Q_\ell)$ (see §4). Then, in fact, we have $V(m) = H^i_c(X,Q_\ell(m))$.

We note that if Φ is a Q_ℓ-sheaf then we can also talk of the twists $\Phi(m)$ of Φ.

§4. Properties of ℓ-adic cohomology with compact support.

As before, X is a scheme which is separated and of finite type over an algebraically closed field K. If X is projective the groups $H^i(X,Q_\ell)$ and $H^i_c(X,Q_\ell)$ coincide. Some of the properties stated below will be stated for torsion abelian sheaves on X. Then by taking the torsion sheaves $\mathbb{Z}/\ell^n\mathbb{Z}$, taking inverse limits and tensoring with Q_ℓ we get the corresponding statements for the $H^i_c(X,Q_\ell)$.

Base change ([63, p. 49; [62], XII, §5)

(5.2) Suppose we have a commutative diagram of schemes

$$\begin{array}{ccc} X & \xleftarrow{g'} & X' \\ {\scriptstyle f}\downarrow & & \downarrow{\scriptstyle f'} \\ S & \xleftarrow[g]{} & S' \end{array}$$

and Φ is a torsion abelian sheaf on X. Then $g^*(R^i f_!)(\Phi) \cong (R^i f'_!)(g'^*)$, for all $i \geq 0$.

In the classical case, the analogous result holds for paracompact spaces (see [29], 4.17.1).

<u>Corollary 5.3</u>. Let $\bar{s}$ be a geometric point of S. Then the stalk $(R^i f_! \Phi)_{\bar{s}}$ at $\bar{s}$ of the direct image sheaf $R^i f_! \Phi$ is isomorphic to $H^i_c(X_{\bar{s}}, \tilde{\Phi})$, where $X_{\bar{s}}$ is the geometric fibre of f over $\bar{s}$, and $\tilde{\Phi}$ is the pullback of Φ to $X_{\bar{s}}$, from the map $X_{\bar{s}} \to X$.

<u>Remark</u>. If $\bar{s}$ is centered at a generic point of X, the theorem follows essentially as it does in the classical case, i.e., by taking smaller and smaller étale neighborhoods of $\bar{s}$. So the theorem has content mainly for non-generic points of S, e.g., closed points.

<u>Finiteness of cohomology</u> ([62], XVII, 5.2.8.1, 5.3.8; [63], p. 84, 2.10)

(5.4) Let Φ be a torsion constructible sheaf on X. Then the groups $H^i_c(X, \Phi)$ are finite and they are zero except when $0 \leq i \leq 2 \dim X$.

The groups $H^i_c(X, \mathbb{Q}_\ell)$ are finite-dimensional vector spaces over $\mathbb{Q}_\ell$.

Leray Spectral Sequence and Grothendieck Spectral Sequence

([63], p. 23; [62] XVII, 5.1.8.1)

Let Φ be a torsion sheaf on X. Let $f: X \to Y$ be a morphism of schemes. Then we have a Leray spectral sequence

$$E_2^{pq} = H_c^p(Y, R^q f_! \Phi) \Rightarrow H_c^{p+q}(X, \Phi).$$

The way we use this is as follows.

(5.5) Suppose that all the fibres of f are isomorphic to a fixed scheme Z such that $H_c^q(Z,\Phi) = 0$ except for $q = q_0$. Then $H_c^p(Y, R^{q_0} f_! \Phi) \cong H_c^{p+q}(X,\Phi)$.

Let $X \xrightarrow{f} Y \xrightarrow{g} Z$ be morphisms of schemes. Then we have a Grothendieck spectral sequence

$$E_2^{pq} = (R^p g_!)(R^q f_! \Phi) \Rightarrow R^{p+q}((gf)_! \Phi).$$

(For the classical case, see [31], p. 299). The way we use this is as follows.

(5.6) If $R^q f_! \Phi = 0$ except for $q = q_0$, then

$$(R^p g_!)(R^{q_0} f_! \Phi) \cong R^{p+q_0}((gf)_! \Phi).$$

In most cases in (5.5) we will be taking $Z = \mathbb{A}^d$, affine space of dimension d. Then we use the following result on the cohomology groups of $\mathbb{A}^d$. Let $\ell \neq p$, where p = char K.

(5.7) $$H_c^i(\mathbb{A}^d, \mathbb{Q}_\ell) = \begin{cases} 0, & i \neq 2d \\ \mathbb{Q}_\ell(-d), & i = 2d. \end{cases}$$

The fact that $H_c^i(\mathbb{A}^d, \mathbb{Q}_\ell) = 0$ for $i \neq 2d$ can be seen, for example, by using the Comparison Theorem (5.13) which enables us in some cases to compute ℓ-adic cohomology groups of varieties by looking at corresponding varieties in characteristic 0 and computing classical cohomology. A reference for the fact that the top cohomology is $Q_\ell(-d)$ is [85], p. 96. We note that it is very important to keep track of the Tate twist here, as this tells us that if F is the Frobenius morphism of $\mathbb{A}^d$, the induced action on $H^i_c(\mathbb{A}^d, Q_\ell)$ is multiplication by q^d.

<u>Functoriality of cohomology groups.</u>

Suppose we have a morphism $f: X \to Y$ of schemes and Φ is a sheaf on Y. As in the case of classical sheaf cohomology, we have an induced morphism $H^i(Y,\Phi) \to H^i(X,f^*\Phi)$. In the compact support case, we have the following. Suppose $f: X \to Y$ is a <u>finite</u> morphism. It can be shown that we can embed X,Y in proper varieties $\tilde{X},\tilde{Y}$ with morphisms $j: X \to \tilde{X}$, $j': Y \to \tilde{Y}$ and $\tilde{f}: \tilde{X} \to \tilde{Y}$ such that $\tilde{f}$ maps $\tilde{X}-X$ into $\tilde{Y}-Y$. Then, by definition, $H^i(\tilde{X}, j_!f^*\Phi) = H^i_c(X,f^*\Phi)$ and $H^i(\tilde{Y}, j'_!\Phi) = H^i_c(Y,\Phi)$. But then by the definitions of $j_!$ and $j'_!$ we have $\tilde{f}^*(j'_!\Phi) = j_!(f^*\Phi)$. Hence we have a homomorphism $H^i_c(Y,\Phi) \to H^i_c(X,f^*\Phi)$.

We will also come across the following situation. Let $f: X \to Y$ be a morphism of schemes. Let g be an automorphism of X (e.g., g belongs to a group acting on X) such that each fibre of f is stable under g. Let Φ be a constant sheaf on X. Then $g^*\Phi$ is a constant sheaf on X, canonically isomorphic to Φ. Thus we have a morphism $g^*\Phi \to \Phi$ which leads to a morphism $\tilde{g}: (R^if_!)(g^*\Phi) \to R^if_!\Phi$ of sheaves on Y. By Base Change (5.2) $(R^if_!)(g^*\Phi)$ is isomorphic to $R^if_!\Phi$. Thus we have an induced action $\tilde{g}$ on the sheaf $(R^if_!\Phi)$ on Y.

More generally, let $g,g!$ be automorphisms of X,X' respectively such that

$$\begin{array}{ccc} X & \xrightarrow{g} & X \\ f\downarrow & & \downarrow f \\ Y & \xrightarrow[g']{} & Y \end{array}$$

is commutative. Let Φ be a constant sheaf on X. Then we have an induced morphism $\tilde{g}\colon g'^{*}(R^i f_!\Phi) \to R^i f_!\Phi$. If $R^i f_!\Phi$ is also a constant sheaf on Y, then we have a morphism of $R^i f_!\Phi$. (As an example, see Theorem 7.3, Chapter VII.)

<u>Long exact sequence</u> ([62], XVII, p. 350, 5.1.16.3)

If we have a short exact sequence of torsion abelian sheaves $0 \to \Phi \to G \to H \to 0$, where Φ,G,H are sheaves on X, we have a long exact sequence of cohomology $0 \to \cdots \to H^{i-1}(X,H) \to H^i(X,\Phi) \to H^i(X,G) \to H^i(X,H) \to H^{i+1}(X,\Phi) \to \cdots$, since the functors $\{R^i\Gamma\}$ form a δ-functor. Now suppose we have a closed subscheme X' of X and let $Y = X - X'$. If Φ is a torsion sheaf on X and $i\colon X' \to X$, $j\colon Y \to X$ are inclusions, then the exact sequence $0 \to j_!\Phi \to \Phi \to i_*(\Phi|Y) \to 0$ of sheaves on X gives rise to a long exact sequence

(5.8) $$\cdots \to H_c^i(Y,\Phi) \to H_c^i(X,\Phi) \to H_c^i(X',\Phi) \to \cdots$$

<u>Künneth formula</u> ([62], XVII, p. 368, 5.4.3)

Let X_1,X_2 be schemes. Then there is an isomorphism

(5.9) $$H_c^k(X_1 \times X_2,Q_\ell) \cong \bigoplus_{i+j=k} \{H_c^i(X_1,Q_\ell) \otimes H_c^j(X_2,Q_\ell)\}.$$

Cohomology of a quotient of a scheme by a finite group.

(5.10) Suppose G is a finite group which acts on a scheme X, and suppose the quotient $Y = X/G$ exists. (This is always the case, for example, if X is quasiprojective; see e.g. [67], p. 59.) Then $H_c^i(Y,Q_\ell) \cong H_c^i(X,Q_\ell)^G$. [Note that G acts on $H_c^i(X,Q_\ell)$; the right hand side denotes the G-invariants in $H_c^i(X,Q_\ell)$.]

There is a classical analogue of this theorem for a locally compact space which has a finite group acting on it and one proves this by means of a transfer map. (See [6], p. 37.) In the étale case also such a transfer map exists ([62], XVII, p. 426, 6.2.5). We will sketch a proof of (5.11) based on [6], loc. cit.

Let Φ be a constant torsion sheaf on Y, then $f^*\Phi$ is a constant sheaf on X and there is a morphism (morphism of adjunction; see e.g. [24], p. 232) $\Phi \to f_!f^*\Phi$ of sheaves on Y. Let $\pi : X \to Y$ be the canonical map. Since each fibre of π is finite we have by (5.5) that $H_c^i(Y,f_!f^*\Phi) \cong H_c^i(X,f^*\Phi)$. Now the action of G on X leads to an action of G on the sheaf $(f_!f^*\Phi)$ on Y. At any geometric point $\bar{y}$ of Y the stalk $(f_!f^*\Phi)_{\bar{y}}$ is a sum of $|G|$ copies of $f^*\Phi$, labeled by the elements of G, and G permutes these abelian groups. Then the group of fixed points under G of this stalk is just $\Phi_{\bar{y}}$ which is embedded in $(f_!f^*\Phi)_{\bar{y}}$. We see that $(f_!f^*\Phi)^G = \Phi$. Now the inclusion $\Phi \to (f_!f^*\Phi)^G$ and the map $(f_!f^*\Phi) \to \Phi$

given by $x \to \sum_{g\varepsilon G} gx$ (the transfer map) give rise to maps

$$\mu\colon H_c^i(X,f^*\Phi) \to H_c^i(Y,f_!f^*\Phi) \to H_c^i(Y,\Phi),$$

$$\sigma\colon H_c^i(Y,\Phi) \to H_c^i(Y,(f_!f^*\Phi)^G) \to H_c^i(Y,f_!f^*\Phi) \to H_c^i(X,f^*\Phi),$$

and we can check that $\mu\sigma$ is multiplication by $|G|$ and $\sigma\mu$ is action by $\sum_{g\varepsilon G} g$. Furthermore σ is just the map $H_c^i(Y,\Phi) \to H_c^i(X,f^*\Phi)$ induced by functoriality form π, and since π commutes with the action of G (regarding G as acting trivially on Y) σ maps $H_c^i(Y,\Phi)$ into $H_c^i(X,f^*\Phi)^G$. Then $\sigma\mu$ is multiplication by $|G|$ on $H_c^i(X,f^*\Phi)^G$.

Now we take $\Phi_n = \mathbf{Z}/\ell^n\mathbf{Z}$, take projective limits of cohomology groups and tensor with Q_ℓ, so that our cohomology groups are then vector spaces over Q_ℓ. We then see that we have an isomorphism $H_c^i(Y,Q_\ell) \xrightarrow{\sim} H_c^i(X,Q_\ell)^G$, as required.

The following two theorems are useful in computing cohomology. An example will be given at the end.

Specialization Theorem ([62], XVI, Theorem 2.1 and Corollary 2.5).

Let R be a discrete valuation ring with residue field k and quotient field $\tilde{k}$. Then Spec R has exactly two points: a closed point x_0 with $k(x_0) = k$ and a generic point x_1 with $k(x_1) = \tilde{k}$ (see [32], p. 74). Let $\bar{x}_j$ be a geometric point of Spec R centered at x_j $(j=0,1)$. Let X be a scheme over R, i.e., let $f\colon X \to \text{Spec } R$ be a morphism, and let Φ be a torsion abelian sheaf on X. By Base Change (5.3) we have that the stalk $(R^i f!)(\Phi)_{\bar{x}_j}$ of $(R^i f!)(\Phi)$ at $\bar{x}_j$ is isomorphic to $H^i_c(X_{\bar{x}_j}, \Phi_j)$ (j=0,1) where Φ_j is the pullback of Φ to $X_{\bar{x}_j}$. Now if in addition we have f proper and smooth (see [32], p. 268 for the definition of a smooth morphism of schemes) and Φ is a p'-torsion abelian sheaf where p = char k, then the Specialization Theorem says that the stalks at both points are isomorphic, i.e.

$H^i_c(X_{\bar{x}_0}, \Phi_0) \cong H^i_c(X_{\bar{x}_1}, \Phi_1)$.

<u>Comparison Theorem</u> ([62], XI, 4.4; [63], p. 51).

(5.13) Let X be a scheme (separated, finite type) over $\mathbb{C}$. Then $H^i_c(X, \mathbb{Z}/n\mathbb{Z}) \cong H^i_c(X^{an}, \mathbb{Z}/n\mathbb{Z})$ for any n, where X^{an} is the space X with its classical topology and the cohomology group on the right is the usual cohomology group with compact support.

<u>The Frobenius morphism</u> ([85], p. 100; [63], p. 79, 80)

Suppose we have a scheme X_0 which is separated and of finite type over $\mathbb{F}_q$. Let $K = \bar{\mathbb{F}}_q$. Then $X = X_0 \times_{\mathbb{F}_q} \operatorname{Spec} K$ is obtained from X_0 by base extension and is a scheme over K.

The Frobenius morphism F_0 of X_0 is the morphism of ringed spaces (X_0, O_{X_0}) which is the identity on the underlying topological space X_0, whereas on any $\alpha \in O_{X_0}(U)$ where U is a (Zariski) open set of X_0, it acts as $\alpha \to \alpha^q$. Since F_0

acts trivially on $(X_0)_{et}$ it induces the identity on $H_c^i(X_0,Q_\ell)$. We now have a morphism $F_0 \times \sigma$ of X, where σ is the element of the Galois group $G(K,\mathbb{F}_q)$ given by $x \to x^q$. Again $F_0 \times \sigma$ induces the identity on $H_c^i(X,Q_\ell)$ and hence $F_0 \times 1$ and $1 \times \sigma$ act as inverses on $H_c^i(X,Q_\ell)$. The morphism $F = F_0 \times 1$ is called the geometric Frobenius morphism and $1 \times \sigma$ is called the arithmetic Frobenius morphism of X.

Remark. Compare the geometric and arithmetic Frobenius morphisms introduced in Chapter I for affine varieties defined over $\mathbb{F}_q$.

Let Φ be any sheaf on X_0. Then F induces an action on the stalk $\Phi_{\bar{x}}$ of Φ at any geometric point of X_0 centered at a closed point x of X_0, as follows. Suppose we have a morphism $\bar{x}$: Spec $\overline{k(x)} \to X_0$ with image x; then the inverse image of Φ is precisely the constant sheaf $\Phi_{\bar{x}}$ on Spec $\overline{k(x)}$. The morphism σ: Spec $\overline{k(x)} \to$ Spec $\overline{k(x)}$ gives rise by functoriality to a morphism of $\Gamma(\text{Spec }\overline{k(x)},\Phi_{\bar{x}})$ onto $\Gamma(\text{Spec }\overline{k(x)}),\sigma^*\Phi_{\bar{x}})$, i.e., an endomorphism of $\Phi_{\bar{x}}$. We define the local Frobenius morphism F_x of $\Phi_{\bar{x}}$ to be the inverse of this.

We now state the important fixed point formula of Grothendieck (see [63], p. 86, Theorem 3.2; [68], §1; [28]).

Trace Formula 5.11. $|X^F| = \sum_{i\geq 0} (-1)^i \operatorname{Tr}(F,H_c^i(X,Q_\ell))$.

Here X^F is the set of closed points of X fixed by F. If F is an affine variety defined over $\mathbb{F}_q$ (as in Chapter I) this formula enables us to compute the number of $\mathbb{F}_q$-rational points of X.

The following theorem ([65], XXI, Theorem 5.2.2), which will be used in Chapter VII, gives a connection between the local Frobenius morphisms and the action of F on cohomology.

Definition. A Q_ℓ-sheaf Φ on X_0 is called integral if, for every geometric point $\bar{x}$ centered at a closed point x of X_0, F_x has eigenvalues on $\Phi_{\bar{x}}$ which are algebraic integers.

Theorem. If Φ is integral, then the eigenvalues of F on $H_c^i(X_0,\Phi)$ are algebraic integers.

Corollary 5.12. Let Φ be a Q_ℓ-sheaf on X_0 such that for all geometric points centered at closed points the eigenvalues of F_x on the stalks are all divisible by q^n. Then the eigenvalues of F on $H_c^i(X_0,\Phi)$ are divisible by q^n.

Proof. Apply the above theorem to $\Phi(-n)$.

Remark. This theorem is proved for $d = 1$ (where $d = \dim X_0$) by using two expressions for the "L-function" $Z(X_0,\Phi,t)$ (see [68], p. 2) and then by induction on d by fibering X_0 over a curve.

We now make some remarks about reductive group schemes. For the definition of a group scheme, see e.g., [32], p. 324. Roughly, a group scheme $\mathfrak{G}$ over $\mathbb{Z}$ is a scheme together with multiplication, identity and inverse morphisms satisfying certain axioms.

Consider the reductive linear algebraic group G defined over $\mathbb{F}_q$ that we have studied in Chapters I through IV. Assume that G is split over $\mathbb{F}_q$ (see 2.11). Grothendieck and Demazure [61] have shown that there is a reductive group scheme $\mathfrak{G}$ over $\mathbb{Z}$ from which we can recover G. In other words, there is a scheme $\mathfrak{G} \to \operatorname{Spec} \mathbb{Z}$ such that we can extend the base and get a morphism $f: \mathfrak{G} \times_{\mathbb{Z}} \operatorname{Spec} A \to \operatorname{Spec} A$ where A is a discrete valuation ring with residue field $\mathbb{F}_q$, the fibre of f over the closed point of $\operatorname{Spec} A$ is isomorphic to G, and the fibre over the generic point of $\operatorname{Spec} A$ is isomorphic to a corresponding group in characteristic 0. Similarly we have a scheme $\mathfrak{G}/\mathfrak{B}$ over $\mathbb{Z}$ which gives rise to the variety G/B where B is a Borel subgroup of G. The Specialization and Comparison theorems enable us to compare the ℓ-adic cohomology of G, G/B, etc., with the classical cohomology of corresponding varieties in characteristic 0. We also make the following remark: we have a morphism $\mathfrak{G}/\mathfrak{B} \to \operatorname{Spec} A$ with the group W (where $W = W(T_0)$ as before) acting on all the fibres. Hence, if Φ is a suitable sheaf on $\mathfrak{G}/\mathfrak{B}$ we can assume that the isomorphisms between the cohomology groups of different fibres with coefficients in Φ given by the Specialization Theorem are W-equivariant.

We now prove a theorem due to Steinberg ([82], 14.14) which will be used in Chapter VI, as an illustration of the theorems mentioned in this chapter.

Theorem 5.13. Let G be a connected reductive algebraic group defined over $\mathbb{F}_q$, and F the Frobenius morphism. Then the number of F-stable maximal tori is $|G^F|_p^2$.

Proof. Let $T_0, B_0, W = W(T_0)$, $N = N(T_0)$ be as in Chapter II. Then we want to compute $|(G/N)^F|$. Now W acts on G/T_0 by the rule $w(xT_0) = xT_0\dot{w}$, and $G/N \cong (G/T_0)/W$. By the Trace Formula (5.11) we have

$$|(G/N)^F| = \sum_i (-1)^i \mathrm{Tr}(F, H_c^i(G/N, Q_\ell))$$

$$= \sum_i (-1)^i \mathrm{Tr}(F, H_c^i(G/T_0, Q_\ell)^W) \quad \text{by (5.10)}.$$

There is a morphism $G/T_0 \to G/B_0$ whose fibres are all isomorphic to U_0, and hence to affine space of dimension d, where $B_0 = T_0U_0$. Thus, by (5.5) and (5.7) we have

$$H_c^i(G/T_0, Q_\ell) \cong (H_c^{i-2d}(G/B_0, Q_\ell))(-d).$$

By the Specialization and Comparison Theorems it follows that $H_c^j(G/B_0, Q_\ell) \cong H_c^j(\tilde{G}/\tilde{B}, \mathbb{C})$ where $\tilde{G}$ is a group analogous to G in characteristic 0 and $\tilde{B}$ is a Borel subgroup of $\tilde{G}$. Moreover, this isomorphism preserves the action of W on cohomology. In characteristic 0 it is known that W acts as the regular representation on $\bigoplus_j H^j(\tilde{G}/\tilde{B}, \mathbb{C})$ and the trivial representation occurs in dimension 2d. (See e.g., I. N. Bernshtein, I. M. Gelfand, and S. I. Gelfand, Schubert cells and flag space cohomologies, Functional Analysis and its Applications, 7 (1973), pp. 53-55). So we have the same situation for $\bigoplus_j H_c^j(G/B_0, Q_\ell)$. Now just as in (5.7) we see that $H_c^{2d}(G/B_0, Q_\ell)$ is isomorphic to $Q_\ell(-d)$ and F acts on this by multiplication by q^d (in fact, this arises from the "cell decomposition" of G/B_0, see e.g., [5], p. 347). Hence on $H_c^{2d}(G/B_0, Q_\ell)(-d)$, F acts as multiplication by q^{2d}. From this it follows that $|(G/N)^F| = q^{2d}$, as required.

CHAPTER VI. THE CONSTRUCTION OF LUSZTIG-DELIGNE

References for this chapter are [18], [48], and [69]. From now on K will be an algebraic closure of $\mathbb{F}_q$.

We first recall the principal series representations of G^F (Chapter III). The subgroups T_0, B_0, U_0 are as in Chapters II and III. If λ is a complex character of T_0^F which is lifted to a character $\tilde{\lambda}$ of B_0^F, then $\mathrm{Ind}_{B_0^F}^{G^F} \tilde{\lambda}$ is a character of G^F which is irreducible if λ is regular. We would now like to generalize this to the case of any torus T^F. In other words, we would like to construct a family of virtual representations (i.e., elements of the Grothendieck group) of G^F parametrized by the characters of T^F and having certain nice properties, e.g., that a virtual representation is irreducible (up to sign) if it corresponds to a regular character of T^F, and that the dimension of each of these virtual representations is $|G^F|/|T^F||U_0^F|$, even though T^F may not be contained in any subgroup of order $|T^F||U_0^F|$. The construction of Lusztig-Deligne achieves exactly this.

Let us view the principal series representations as being constituents of $\mathrm{Ind}_{U_0^F}^{G^F}(1)$. Thus we can construct them by taking the space G^F/U_0^F, and letting T_0^F act on the right and G^F on the left on this space. We can decompose the space by means of the characters of T_0^F and each subspace corresponding to a fixed character of T_0^F is G^F-stable. Thus we get representations of G^F indexed by the characters of T_0^F. Now we have $G^F/U_0^F \cong (G/U_0)^F$ by Lemma 2.6. Thus G^F/U_0^F is isomorphic to the subvariety of the variety G/U_0 given by $\{gU_0 | g \in G,\ g^{-1}(Fg) \in U_0\}$.

Now suppose T is any F-stable maximal torus. Let $B = TU$ where the Borel subgroup B (and hence U) is not necessarily F-stable. We define the <u>Lang covering</u> $L^{-1}(U)$ of U as follows.

<u>Definition</u>. $L^{-1}(U) = \{g \in G | g^{-1}(Fg) \in U\}$.

Then $L^{-1}(U)$ is a closed subset of G and hence is an affine variety. We see that $G^F \times T^F$ acts on $L^{-1}(U)$ by $(g,t)h = ght^{-1}$ $(g \in G^F,\ t \in T^F,\ h \in L^{-1}(U))$. Now in order to exploit this fact and get a linear representation of $G^F \times T^F$ in characteristic 0 we go over to the ℓ-adic cohomology of $L^{-1}(U)$. <u>Let ℓ be a prime different from p</u>. By functoriality we get an action of $G^F \times T^F$ on the cohomology groups $H_c^i(L^{-1}(U),\overline{\mathbb{Q}}_\ell)$. From now on we let $\hat{T}^F$ denote the group of characters $\mathrm{Hom}(T^F,\overline{\mathbb{Q}}_\ell^*)$. For any $\theta \in \hat{T}^F$, and any T^F-module M over $\overline{\mathbb{Q}}_\ell$ let M_θ denote the θ-isotypic part of M. We see that the subspace $H_c^i(L^{-1}(U),\overline{\mathbb{Q}}_\ell)_\theta$ of $H_c^i(L^{-1}(U),\overline{\mathbb{Q}}_\ell)$ is G^F-stable.

<u>Definition</u>. $R_T^G(\theta)$ is the virtual representation $\sum_{i \geq 0} (-1)^i H_c^i(L^{-1}(U),\overline{\mathbb{Q}}_\ell)_\theta$ of G^F.

Thus we have a map $\theta \to R_T^G(\theta)$ from $\hat{T}_F$ into the Grothendieck group $\mathcal{R}(G^F)$ of G^F. It will be shown later (6.18) that $R_T^G(\theta)$ depends only on T and not on the choice of U, and so we are justified in omitting U from the notation. We will also denote the variety $L^{-1}(U)$ by $\tilde{X}$.

Example. (See [48], p. 17.) $G = SL_2$, $G^F = SL(2,q)$. Let T^F be a "non-split" torus of G^F of order $q+1$ (see Chapter II for the case of GL_2). Then T^F is isomorphic to the group of elements of norm 1 in $\mathbb{F}^*_{q^2}$. It can be checked that $\tilde{X} = L^{-1}(U)$ is the affine curve $xy^q - yx^q = 1$ on which G^F acts according to its natural action on $(x,y) \in K^2$ and T^F acts as $\lambda(x,y) = (\lambda x, \lambda y)$ $(\lambda \in \mathbb{F}^*_{q^2})$. Then $H^0_c(\tilde{X},\overline{\mathbb{Q}}_\ell) = 0$, $\dim H^2_c(\tilde{X},\overline{\mathbb{Q}}_\ell) = 1$, and T^F acts trivially on this space. If $\theta \in \hat{T}^F$ is not trivial, then $-R^G_T(\theta)$ is realized as the θ-isotypic component of $H^1_c(\tilde{X},\overline{\mathbb{Q}}_\ell)$. If $\theta^2 \neq 1$, $-R^G_T(\theta)$ is irreducible of dimension $q - 1$ and this is the family of discrete series representations. Let $\theta^2 = 1$, $\theta \neq 1$. The θ-component on $H^1_c(\tilde{X},\overline{\mathbb{Q}}_\ell)$ then splits into two representations of G^F of dimension $\frac{1}{2}(q-1)$.

We briefly sketch how the dimensions of the cohomology groups can be computed. We embed the affine curve A defined by $xy^q - yx^q = 1$ in the non-singular projective curve C given by $xy^q - yx^q - z^{q+1} = 0$ in the projective space $\mathbb{P}^2$ over K. A is the open subset of C given by $z \neq 0$. We then use the long exact sequence (5.8) with respect to C, A, and $C - A$.

By a vanishing theorem for cohomology of affine schemes [63], p. 51) we see that $H^0_c(A,\overline{\mathbb{Q}}_\ell) = 0$. Since C is a projective non-singular curve, we get $\dim H_c^0(C,\bar{Q}_\ell) = \dim H_c^2(C,\bar{Q}_\ell) = 1$ and $\dim H_c^1(C,\bar{Q}_\ell) = 2g$ where g is the genus of C (see [63], p. 35), and in this case $g = q\frac{(q-1)}{2}$. [The groups given there are the $H^i(C,\bar{Q}_\ell)$. To get the $H_c^i(C,\bar{Q}_\ell)$, we use duality; see [63], p. 71.] From these facts and the fact that $\dim H_c^0(C-A,\bar{Q}_\ell) = 1$, $\dim H_c^i(C-A,\bar{Q}_\ell) = 0$ $(i=1,2)$, we get that $\dim H_c^1(A,\bar{Q}_\ell) = q(q-1)$ and $\dim H_c^2(A,\bar{Q}_\ell) = 1$.

We also remark (see loc. cit.) that the two representations of dimension $\frac{1}{2}(q-1)$ (the so-called Hecke representations of $SL(2,q)$) can be realized in $H_c^1(C',\bar{Q}_\ell)$ where C' is the hyperelliptic curve $y^2 = x^q - x$, on which $SL(2,q)$ acts. Allan Adler has pointed out to me that this is a curve which in characteristic p provides a counterexample to the theorem of Hurwitz that a curve of genus $g \geq 2$ over a field of characteristic 0 has at most $84(g-1)$ automorphisms (see [32], p. 306).

For other examples in classical groups, see [18], p. 116. The variety $\tilde{X}(w)$ mentioned there is a quotient of our variety $\tilde{X}$, if our torus T corresponds to $w \in W(T_0)$ in the sense of Chapter II, 2.9. The cohomology of $\tilde{X}(w)$ is the same as that of our $\tilde{X}$.

Before developing the main properties of the $R_T^G(\theta)$ we discuss some general results on schemes which have automorphisms of finite order, and on the induced morphisms on cohomology. As in Chapter V, let X be a scheme which is separated and of finite type over K. Let g be an automorphism of X of finite order. We then have an action of g on $H_c^i(X,\overline{Q}_\ell)$. From now on we will always denote $H_c^i(X,\overline{Q}_\ell)$ by simply $H_c^i(X)$ when there is no danger of ambiguity. We now define the <u>Lefchetz number</u> $\mathcal{L}(g,X)$ of g as the alternating trace of g on the cohomology of X.

<u>Definition</u>. $\mathcal{L}(g,X) = \sum_i (-1)^i \operatorname{Tr}(g,H_c^i(X))$.

We have the following additivity property of the Lefchetz number.

(6.1) Suppose X is a finite disjoint union of locally closed subschemes $\{X_i\}$ which are stable under g. Then $\mathcal{L}(g,X) = \sum_i \mathcal{L}(g,X_i)$. This can be proved using the Long Exact Sequence (5.8).

We remark that X can be written as a finite disjoint union of locally closed quasiprojective schemes which are stable under g. (This follows from the fact that X is of finite type and g is of finite order.)

<u>Proposition 6.2</u> ([18], 3.3; [69], Proposition 1).

$\mathcal{L}(g,X)$ is an integer which is independent of ℓ.

<u>Proof</u>. By (6.1) and the above remark we can assume that X is quasiprojective. We can also assume that X is defined over F_q for some q, and that $F: X \to X$ is the associated Frobenius morphism. Then by (2.1) we have that $F^n g$ is also a Frobenius morphism for some way of defining X over F_q. By the trace formula (5.11) we then have $\mathcal{L}(F^n g,X) = |X^{F^n g}|$. Thus $|X^{F^n g}| = \sum_\lambda a_\lambda \lambda^n$ where a_λ is the alternating trace of g

on the generalized λ-eigenspace of F on $\bigoplus_i H_c^i(X)$ and λ runs over the eigenvalues of F on $\bigoplus_i H_c^i(X)$. Thus $\sum_\lambda a_\lambda \lambda^n$ is an integer which is independent of ℓ.

Now the functions $n \to \lambda^n$ from $\mathbb{Z}^+ \to \overline{\mathbb{Q}}_\ell^*$ are linearly independent; this can be seen, for example, by extending them to characters of $\mathbb{Z}$, and using Dedekind's theorem. Let $\sigma \in G(\overline{\mathbb{Q}}_\ell, \mathbb{Q})$. Then $\sigma(\sum a_\lambda \lambda^n) = \sum a_\lambda \lambda^n$. On the other hand, $\sum \sigma(a_\lambda)\sigma(\lambda)^n = \sum \sigma(a_{\sigma^{-1}(\lambda)})\lambda^n$, and thus $\sigma^{-1}(a_\lambda) = a_{\sigma^{-1}(\lambda)}$ for any σ. Thus $\sum_\lambda a_\lambda \in \mathbb{Q}$, but this is just $\mathcal{L}(g,X)$. But $\mathcal{L}(g,X)$ is a character value of the finite group $\langle g\rangle$ and so it is an algebraic integer. Thus $\mathcal{L}(g,X)$ is an integer.

To show that $\mathcal{L}(g,X)$ does not depend on ℓ, let us denote for the moment by $\mathcal{L}(g,X)_\ell$ the alternating trace of g on $\bigoplus_i H_c^i(X,\overline{\mathbb{Q}}_\ell)$. Let ℓ' be another prime ($\neq p$) and let $\tau\colon \overline{\mathbb{Q}}_\ell \to \overline{\mathbb{Q}}_{\ell'}$ be an isomorphism. The same argument as above shows that $\tau(\mathcal{L}(g,X)_\ell) = (\mathcal{L}(g,X))_{\ell'}$, and this proves the result.

We now state an important theorem, one of the central theorems in the work of Lusztig-Deligne ([18], Theorem 3.2). We will give the main ideas of the proof but some of the details will be omitted.

<u>Theorem 6.3</u>. Suppose the automorphism g of finite order of the scheme X can be written as $g = su = us$ where u is of order a power of p and s is of order prime to p. Then $\mathcal{L}(g,X) = \mathcal{L}(u,X^s)$, where X^s is the subscheme of X of fixed points of s.

Sketch of proof. As in (6.2) we can assume that X is quasiprojective. Furthermore we can assume that the cyclic group $\langle g\rangle$ acts freely on X. (For we can write X as a disjoint union of subschemes X_i where each element of X_i has the same stabilizer H_i. Then we can consider the quotient scheme X_i/H_i.)

Now suppose that g is a p-element. Then $s=1$, $u=g$ and there is nothing to prove. So we can assume that the order of g is divisible by some prime different from p, and since $\mathcal{L}(g,X)$ is independent of ℓ we can assume that this prime is ℓ. Since s has no fixed points on X we have to show that $\mathcal{L}(g,X) = 0$. Thus the theorem will follow from the following proposition.

(6.4) If H is a finite group acting freely on X, the function $h \to \mathcal{L}(h,X)$ $(h \in H)$ is the character of a virtual projective $Z_\ell[H]$-module.

The theorem then follows from (6.4) since it is well-known that the character of the representation of H on a projective $Z_\ell[H]$-module vanishes on elements of order divisible by ℓ (see e.g., [66], p. 133).

We sketch a proof of (6.4). Let $\pi: X \to Y$ be the natural map, where $Y = X/H$. (Since X is quasiprojective, the quotient exists.) Since the fibres of π are finite, by (5.5) we have $H_c^i(X,\mathbb{Z}/\ell^n\mathbb{Z}) \cong H_c^i(Y,\pi_!\mathbb{Z}/\ell^n\mathbb{Z})$. Let $A = (\mathbb{Z}/\ell^n\mathbb{Z})[H]$, so that A is a noetherian ring. An argument similar to that of (5.10) shows that $\pi_!\mathbb{Z}/\ell^n\mathbb{Z}$ is a sheaf of free A-modules of rank 1 on Y. However, the difficulty now is that the $H^i_c(Y,\pi_!\mathbb{Z}/\ell^n\mathbb{Z})$ need not be free or projective A-modules. Thus it does not make sense to talk of the trace of an element of H on the $H^i_c(Y,\pi_!\mathbb{Z}/\ell^n\mathbb{Z})$. However, since A is noetherian, they are finitely generated A-modules and they vanish unless $0 \leq i \leq 2 \dim Y$ (see [62], XVII, 5.2.8.1 for the vanishing and 5.3.6 for the finiteness).

We can compute the $H^i_c(Y,\pi_!\mathbb{Z}/\ell^n\mathbb{Z})$ by taking a suitable resolution of $\pi_!\mathbb{Z}/\ell^n\mathbb{Z}$. We could take the Godemont resolution by a complex $\dot{\tau}: \tau^0 \to \tau^1 \to \tau^2 \to \cdots$ of flasque sheaves (see [29], p. 167 for the classical case and [62], XVII, 4.2 for the étale case). This is defined by taking, for any U with $U \to Y$, $\tau^0(U) = \Pi(\pi_!\mathbb{Z}/\ell^n\mathbb{Z})_{\bar{x}}$ where the product is over geometric points $\bar{x}$ of U. Then we proceed analogously to construct $\tau^1, \tau^2, \ldots$ and so on. By taking global sections we get a complex $\dot{C}: C^0 \to C^1 \to C^2 \to \cdots$ of A-modules which are flat A-modules in our case, since each stalk is a free A-module and the direct product of flat A-modules is flat if A is noetherian (see e.g., [24], pp. 439, 440). The cohomology groups of the complex $\Gamma(\pi_!\mathbb{Z}/\ell^n\mathbb{Z}) \to C^0 \to C^1 \to \cdots$ are the groups $H^i_c(Y,\pi_!\mathbb{Z}/\ell^n\mathbb{Z})$. The A-modules C^i need not be finitely generated; however, we can replace $\dot{C}$ by another complex with the same cohomology such that the terms of the complex are finitely generated flat A-modules. This is guaranteed by the following lemma. A proof of this lemma is given in [55], p. 47 (see also [32], p. 283).

Lemma. Let A be a noetherian ring. Let $\dot{C}$ be a complex of A-modules such that the $H^i(\dot{C})$ are finitely generated A-modules and such that $C^p \neq 0$ only if $0 \le p \le m$. Then there is a complex $\dot{K}$ of finitely generated A-modules such that $K^p \neq 0$ only if $0 \le p \le m$ and K^p is free if $1 \le p \le m$, and a homomorphism of complexes $\phi: \dot{K} \to \dot{C}$ such that ϕ induces isomorphisms $H^i(\dot{K}) \tilde{\to} H^i(\dot{C})$ for all i. Moreover, if the C^p are A-flat, then K^0 is also A-flat.

Applying this to our situation we find that the $H^i_c(Y, \pi! \mathbf{Z}/\ell^n \mathbf{Z})$ are the cohomology groups of a complex $\dot{K}_n$ such that (i) K^p_n are finitely generated A-modules which are zero unless $0 \le p \le 2 \dim Y$ (ii) K^p_n is free if $p \ge 1$ (iii) K^0_n is A-flat and finitely generated, hence projective.

We need one more fact, namely that once the complex $\dot{K}_n$ is chosen, we can choose $\dot{K}_{n+1}$ in such a way that $\dot{K}_{n+1}$ is the reduction $\mod \ell^n$ of $\dot{K}_n$. We omite the proof of this. (See [63], p. 97, 4.12, and [64], XV, p. 473.) Then, taking projective limits we get a complex $\dot{K}_\infty$ of projective A-modules such that $\lim_{\leftarrow} H^i_c(Y, \pi! \mathbf{Z}/\ell^n \mathbf{Z}) = H^i(\dot{K}_\infty)$, and thus $H^i_c(X, \mathbf{Z}_\ell) \cong H^i(\dot{K}_\infty)$ since $H^i_c(X, \mathbf{Z}/\ell^n \mathbf{Z}) \cong H^i_c(Y, \pi! \mathbf{Z}/\ell^n \mathbf{Z})$ for each n. Moreover, this isomorphism is H-equivariant.

Let $h \in H$. We now see that $\mathcal{L}(h, X)$, the alternating trace of h on $\oplus H^i_c(X, Q_\ell)$, is equal to the alternating trace of h on the complex $\dot{K}_\infty$ of projective A-modules. This proves (6.4) and hence Theorem 6.3 .

Remark. The proof in [18], Theorem 3.2 uses the language of derived categories, which we have avoided in order not to introduce too much machinery. However, it may be worthwhile making a few remarks about them as they are a valuable tool in the theory; for example, Deligne's proof of the Trace Formula in [63] makes essential use of them. There, also, we run into the problem that certain cohomology groups are not free modules.

Let $\mathfrak{A}$ be an abelian category. The derived category $D\mathfrak{A}$ is constructed in two steps: (i) Take the category whose objects are homotopy equivalence classes of chain complexes of objects of $\mathfrak{A}$ and morphisms are the obvious morphisms of chain complexes; (ii) "Invert" all quasi-isomorphisms (i.e., morphisms of complexes which induce isomorphisms of cohomology). This is done by the "Calculus of Fractions" and is roughly analogous to constructing a quotient ring of a ring with respect to a multiplicative subset.

So we can think of $D\mathfrak{A}$ as being the category of homotopy classes of complexes of objects of $\mathfrak{A}$, such that a map $\dot{C}_1 \to \dot{C}_2$ of complexes which induces an isomorphism of cohomology, is an isomorphism in $D\mathfrak{A}$. If $F: \mathfrak{A} \to \mathfrak{B}$ is a left exact functor from an abelian category $\mathfrak{A}$ to another abelian category $\mathfrak{B}$ then the derived functor RF can be defined as a functor from $D\mathfrak{A}$ to $D\mathfrak{B}$. The use of derived categories makes it easier to talk of cohomology, derived functors, spectral sequences, etc. (see [33]).

We now prove a theorem on trivial action of a connected

group on cohomology, which will be used at several points in this and the next chapters.

<u>Theorem 6.5</u>. ([18], Theorem 6.4)

Let H be a connected algebraic group acting on a scheme Y, separated and of finite type over K. Then, if $h \in H$, the action of h on $H_c^i(Y,\mathbf{Z}/\ell^n\mathbf{Z})$ (and hence on $H_c^i(Y,Q_\ell)$) is trivial.

<u>Proof</u>. Let $f\colon H \times Y \to H \times Y$ be the morphism defined by $f(h,y) = (h,hy)$ $(h \in H,\ y \in Y)$ and π the projection of $H \times Y$ on H. We have the following commutative triangle.

$$\begin{array}{ccc} H\times Y & \xrightarrow{f} & H\times Y \\ & {}_{\pi}\searrow \quad \swarrow_{\pi} & \\ & H & \end{array}$$

We apply Base Change ((5.2) and (5.3)) to the commutative diagram

$$\begin{array}{ccc} H\times Y & \longrightarrow & Y \\ {\scriptstyle\pi}\downarrow & & \downarrow \\ H & \longrightarrow & \operatorname{Spec} K \end{array},$$

where the top arrow is projection. Then we see that $R^i\pi_!(\mathbf{Z}/\ell^n\mathbf{Z})$ is the constant sheaf $H_c^i(Y,\mathbf{Z}/\ell^n\mathbf{Z})$ on H. Since f preserves the fibres of π we get an induced morphism, also denoted by f, of this constant sheaf on H. The action of H on Y gives rise to an action of H on $H_c^i(Y,\mathbf{Z}/\ell^n\mathbf{Z})$. Since $f(h,y) = (h,hy)$, f acts on the stalk of the constant sheaf $H_c^i(Y,\mathbf{Z}/\ell^n\mathbf{Z})$ at a geometric point centered at $h \in H$ in exactly the same way as the action of h on $H_c^i(Y,\mathbf{Z}/\ell^n\mathbf{Z})$

induced from the action of h on Y. If $h = 1$, the action is trivial. We have a sheaf of endomorphisms of the constant sheaf $H^i_c(Y,\mathbb{Z}/\ell^n\mathbb{Z})$ on the connected space H, and this itself is a constant sheaf. This means that the action of f is trivial at all $h \in H$, as required.

Remark 6.6. We are looking at closed points h of H and K is algebraically closed. Thus we can think of the fibre of π over a geometric point centered at h as just being isomorphic to Y; cf. the remark below our discussion of geometric points.

We now return to the virtual representations $R_T^G(\theta)$, corresponding to an F-stable maximal torus T of G and $\theta \in \hat{T}$, which are realized on the cohomology of a scheme $\tilde{X} = L^{-1}U$, where TU is a Borel subgroup of G. The rest of this chapter will be devoted to proving various properties of the $R_T^G(\theta)$.

We first show that if $u \in G^F$ is unipotent, then $\mathrm{Tr}(u,R_T^G(\theta))$ is independent of θ. For,

$$\begin{aligned}\mathrm{Tr}(u,R_T^G(\theta)) &= \frac{1}{|T^F|}\sum_{t\in T^F}\mathcal{L}((u,t),\tilde{X})\,\theta(t)\\ &= \frac{1}{|T^F|}\sum_t \mathcal{L}(u,\tilde{X}^t)\,\theta(t), \text{ from (6.3)}\\ &= \frac{1}{|T^F|}\mathcal{L}(u,\tilde{X}), \text{ since } t \text{ acts}\end{aligned}$$

fixed point freely on $\tilde{X}$ if $t \neq 1$. By (6.2), $\mathcal{L}(u,\tilde{X})$ is an integer and thus $\mathrm{Tr}(u,R_T^G(\theta))$ is rational. Since it is an algebraic integer, it is an integer independent of θ.

<u>Definition 6.7.</u> $Q_T^G(u) = \mathrm{Tr}(u, R_T^G(1))$.

The function Q_T^G on the unipotent elements of G^F is called a <u>Green function</u>, after Green [30] who studied them in the case of GL_n.

We now prove an important character formula ([18], 4.2). We first state some facts about centralizers of semisimple elements in G^F (see [7], E-35, E-38).

Let $s \in G^F$, and let s be semisimple. Then $C^0(s)$ is a connected reductive group which is generated by a maximal torus T such that $s \in T$ and all root subgroups U_α with respect to T such that $\alpha(s) = 1$. (For example, this follows from the Bruhat decomposition.) If $B = TU$ (where U is generated by the U_α with α positive), then $B \cap C^0(s)$ is a Borel subgroup of $C^0(s)$ with unipotent radical $U \cap C^0(s)$. Any unipotent element in $C(s)$ lies in $C^0(s)$.

<u>Example.</u> $G = GL_3$,

$$s = \begin{pmatrix} a & 0 & 0 \\ 0 & a & 0 \\ 0 & 0 & b \end{pmatrix} \quad \text{where } a,b \in \mathbb{F}_q.$$

Then $C(s) = C^0(s)$ and $C(s)$ is generated by the diagonal torus T_0 and the root subgroup

$$U_\alpha = \left\{ \begin{pmatrix} 1 & t & 0 \\ 0 & 1 & 0 \\ 0 & 0 & 1 \end{pmatrix} \;\middle|\; t \in K \right\}.$$

<u>Theorem 6.8</u> (Character Formula). Let $g \in G^F$, and let $g = su$ be the Jordan decomposition of g. Then

$$(6.9) \quad \mathrm{Tr}(g, R_T^G(\theta)) = \frac{1}{|C^0(s)^F|} \sum_{\substack{x \in G^F \\ xsx^{-1} \in T^F}} Q^{C^0(s)}_{x^{-1}Tx}(u)\ \theta(xsx^{-1}).$$

Proof. First we note that the formula on the right hand side makes sense since $xsx^{-1} \in T^F$ implies $s \in x^{-1}Tx$ and thus $x^{-1}Tx \subset C^0(s)$; however, T and $x^{-1}Tx$ need not be conjugate in $C^0(s)$. Since $C^0(s)$ is a reductive group we can talk of a virtual representation $R^{C^0(s)}_{x^{-1}Tx}(\theta)$ (and hence a Green function $Q^{C^0(s)}_{x^{-1}Tx}$) of $C^0(s)^F$.

The left hand side of (6.9) is

$$\frac{1}{|T^F|} \sum_{t \in T^F} \mathcal{L}((su,t),\tilde{X})\,\theta(t)$$

$$= \frac{1}{|T^F|} \sum_{t \in T^F} \mathcal{L}(u,\tilde{X}^{(s,t)})\,\theta(t) \quad \text{by (6.3).}$$

Consider, for a fixed t, the scheme $\tilde{X}^{(s,t)}$ $= \{g' \in G \mid g'^{-1}(Fg') \in U,\ sg't^{-1} = g'\}$. Let $g' \in \tilde{X}^{(s,t)}$. Suppose $g'^{-1}(Fg') = v$, so that $Fg' = g'v$. Since $sg't^{-1}$ $= g'$ and $s,t \in G^F$ we get $s(Fg')t^{-1} = Fg' = s(g'v)t^{-1}$ $= g'tvt^{-1}$. Thus $g'v = g'tvt^{-1}$ and $v \in C(t)$. Since $v \in U$, $v \in C^0(t)$. By Lang's Theorem (2.4) we can write $v = z^{-1}(Fz)$ where $z \in C^0(t)$. Now we see that $g'z^{-1} \in G^F$, for $(Fg')(Fz)^{-1}$ $= g'v(zv)^{-1} = g'z^{-1}$. Thus $g' = hz$ where $h \in G^F$, $z \in C^0(t)$.

Next we note that $hz = g' = sg't^{-1} = shzt^{-1} = sht^{-1}z$ and thus $h = sht^{-1}$. Now consider the map

$$\{h\varepsilon G^F \mid h^{-1}sh = t\} \times \{z\varepsilon C^0(t) \mid z^{-1}(Fz) \ \varepsilon\ U \cap C^0(t)\} \to \tilde{X}^{(s,t)}$$

given by $$(h,z) \to hz.$$

This map is surjective by the above remarks. Also, if $h_1z_1 = h_2z_2$ then $h_2^{-1}h_1 = z_2z_1^{-1} \ \varepsilon\ G^F \cap C^0(t) = C^0(t)^F$. Hence if we factor out the product on the left hand side by the action of $C^0(t)^F$ (which acts on the first factor on the right, and on the second factor, which we denote by $\tilde{Y}_t$, on the left) we see that the map

$$\{h \ \varepsilon\ G^F \mid h^{-1}sh = t\} \underset{C^0(t)^F}{\times} \tilde{Y}_t \to \tilde{X}^{(s,t)}$$

is a bijection. Thus we see that we have a partition of $\tilde{X}^{(s,t)}$ into a finite number of subschemes $\{h\} \times \tilde{Y}_t$, where, on $\{h\} \times \tilde{Y}_t$, u acts as $hz \to uhz = h(h^{-1}uh)z$. We also note that $\tilde{Y}_t$ is analogous to $\tilde{X}$, for the group $C^0(t)$. We then get

$$\mathcal{L}(u,\tilde{X}^{(s,t)}) = \frac{1}{|C^0(t)^F|} \sum_{\substack{h\varepsilon G^F \\ h^{-1}sh=t}} \mathcal{L}(h^{-1}uh,\tilde{Y}_t) \qquad \text{(using 6.1))}$$

$$= \frac{1}{|C^0(t)^F|} \sum_{\substack{h\varepsilon G^F \\ h^{-1}sh=t}} |T^F| Q_T^{C^0(t)}(h^{-1}uh).$$

Thus the left hand side of (6.9) becomes

$$\frac{1}{C^0(s)^F} \sum_{\substack{h \in G^F \\ h^{-1}sh \in T^F}} Q^{C^0(s)}_{hTh^{-1}}(u)\ \theta(h^{-1}sh)$$

which is the right hand side of (6.9). This proves the theorem.

Remark. This character formula enables us to express the value of the character of $R_T^G(\theta)$ at g, in terms of functions on the semisimple elements and on the unipotent elements. It is not possible to compute the $Q_T^G(u)$ explicitly from the definition of $R_T^G(\theta)$ and even the computation of $Q_T^G(1)$ (i.e., the dimension of $R_T^G(\theta)$) involves the existence and properties of the Steinberg character. The problem of describing the Green functions $Q_T^G(u)$ will be studied in Chapter VII.

Fom now on, for any scheme X, we will denote $H_c^i(X,\overline{Q}_\ell)$ by $H_c^i(X)$.

Our aim now is to prove an orthogonality theorem for the $R_T^G(\theta)$. We first introduce certain definitions. Recall (Chapter II) that for any F-stable maximal torus T, $X(T)$ denotes the group of characters of T. Let $Y(T) = \mathrm{Hom}\,(K^*,T)$. Then $Y(T)$ is called the group of one-parameter subgroups of T. Both $X(T)$ and $Y(T)$ are free abelian groups of rank ℓ where ℓ is the dimension of T and we have a non-singular pairing $X(T) \times Y(T) \to \mathbf{Z}$ given by $(\chi,\phi) = n$ if $\chi\phi(t) = t^n$ $(t \in K)$ which make them dual to each other (see [5], §8). Now F acts on $X(T)$ and hence on $Y(T) = \mathrm{Hom}(X(T),\mathbf{Z})$. There is a positive integer n such that F^n acts on T as $t \to t^{q^n}$;

this follows from the fact that we can assume T in diagonal form with some action of F on it, and if A is the affine coordinate ring of T there is an n such that the morphism F^n of A takes a to a^{q^n} for each a (see the proof of Theorem 2.10). Then T is split over $\mathbb{F}_{q^n}$.

We now choose a fixed generator γ of the group of $(q - 1)^{st}$ roots of unity in K^*. If T is split over $\mathbb{F}_q$, we have an exact sequence

$$(6.9) \qquad 0 \to Y(T) \overset{F-1}{\to} Y(T) \to T^F \to 0,$$

given as follows: the map $Y(T) \to T^F$ is given by $\phi \to \phi(\gamma)$. We note that $\phi(\gamma) \in T^F$ since $F(\phi(\gamma)) = \phi(\gamma)^q = \phi(\gamma^q) = \phi(\gamma)$. The kernel of this map is $(F - 1)Y(T)$. The map is surjective since it is known that the images of the one-parameter subgroups generate T, i.e., the map $Y(T) \otimes K^* \to T$ given by $(\phi, x) \to \phi(x)$ is an isomorphism.

If T is split over $\mathbb{F}_{q^n}$, we take the map $Y(T) \to T^{F^n}$ as above and compose it with the "norm" map $N = 1 + F + \cdots + F^{n-1}$: $T^{F^n} \to T^F$. Note that we have a commutative diagram

$$(6.10) \qquad \begin{array}{ccccc} Y(T) & \xrightarrow{F^n-1} & Y(T) & \longrightarrow & T^{F^n} \\ N\big\downarrow & & \| & & \big\downarrow N \\ Y(T) & \xrightarrow[F-1]{} & Y(T) & \longrightarrow & T^F \end{array}$$

So in any case we have an exact sequence of the form (6.9) and thus we can regard any $\theta \in \hat{T}^F$ as a character (i.e., a homomorphism $Y(T) \to \overline{\mathbb{Q}}_\ell^*$) of $Y(T)$.

Lemma 6.11 ([18], 5.4). Let $\theta \in \hat{T}^F$, $\theta' \in \hat{T}'^F$, where T, T' are F-stable maximal tori. Then the following conditions are equivalent.

(i) ${}^xT = T'$ for some $x \in G$, such that the induced map $Y(T) \to Y(T')$ takes θ to θ'.

(ii) For some n, there exists $x \in G^{F^n}$ such that ${}^xT^{F^n} = T'^{F^n}$ and ${}^x(\theta\cdot N) = \theta'\cdot N$.

Proof. From the commutative diagram (6.10) it is clear that if $\theta \in \hat{T}^{F^n}$, then θ and $\theta\cdot N \in \hat{T}^F$ give rise to the same character of $Y(T)$ and so (i) is not changed if we replace θ by $\theta\cdot N$. So we can assume T, T' split over $\mathbb{F}_q$, in which case the lemma is clear.

Definition ([18], 5.5). The pairs (T,θ), (T',θ') are said to be geometrically conjugate if either of the two above conditions holds.

Example. $G = GL_2$, F the standard Frobenius. We have two subgroups:

$$T_1 = T_0^F = \left\{\begin{pmatrix}\gamma^a & 0\\ 0 & \gamma^b\end{pmatrix}\right\} \text{ and } T_2 = \left\{\begin{pmatrix}\zeta^a & 0\\ 0 & \zeta^{aq}\end{pmatrix}\right\} \text{ (where } \zeta^{q+1} = \gamma)$$

such that $T_1 \subset G^F$ and T_2 is conjugate in G to a subgroup of G^F. If $S = T_0^{F^2} = \left\{\begin{pmatrix}\zeta^a & 0\\ 0 & \zeta^b\end{pmatrix}\right\}$, we have two norm maps $N_1\colon S \to T_1$, $N_2\colon S \to T_2$ given by $N_1\colon \begin{pmatrix}\zeta^a & 0\\ 0 & \zeta^b\end{pmatrix} \to \begin{pmatrix}\gamma^a & 0\\ 0 & \gamma^b\end{pmatrix}$

and N_2: $\begin{pmatrix} \zeta^a & 0 \\ 0 & \zeta^b \end{pmatrix} \to \begin{pmatrix} \zeta^{a+bq} & 0 \\ 0 & \zeta^{b+aq} \end{pmatrix}$. (For N_2, we use the twisted Frobenius map $\begin{pmatrix} \zeta^a & 0 \\ 0 & \zeta^b \end{pmatrix} \to \begin{pmatrix} \zeta^{bq} & 0 \\ 0 & \zeta^{aq} \end{pmatrix}$ of S.)

Any character of T_1 is given by $\theta_{m,n}$: $\begin{pmatrix} \gamma^a & 0 \\ 0 & \gamma^b \end{pmatrix} \to \gamma^{ma+nb}$

(γ also denotes a primitive (q-1)st root of unity in $\overline{\mathbb{Q}}_\ell^*$ here)

and any character of T_2 is given by τ_n: $\begin{pmatrix} \zeta^a & 0 \\ 0 & \zeta^{aq} \end{pmatrix} \to \zeta^{na}$

(see also Chapter III). Then the characters $\theta_{m,m}$ of T_1 and $\tau_{m(q+1)}$ of T_2 are geometrically conjugate. In particular, $1 \in \hat{T}_1$, $1 \in \hat{T}_2$ are geometrically conjugate.

<u>Definition</u>. $N_G(T,T') = \{g \in G | g^{-1}Tg = T'\}$ and $W_G(T,T')$ is the orbit set $T\backslash N(T,T') \cong N(T,T')/T'$. We also write $W(T,T')$ for $W_G(T,T')$, and $W(T)$ for $W(T,T)$ (which agrees with our previous notation).

We have a variant of the Bruhat decomposition for G. Choose representatives $\dot{w} \in N(T,T')$ for the elements $w \in W(T,T')$. Any element $g \in G$ can be written as $g = u_g n_g u'_g$, where $u \in U \cap n_g U'^- n_g^{-1}$, $n_g \in N(T,T')$, $u'_g \in U'$. We have $G = \bigcup_{w \in W(T,T')} G_w$, where G_w is the set of all g such that $n_g \in T\dot{w}$.

We also note that F acts on $W(T,T')$ and we have

$W(T,T')^F \cong T^F \backslash N(T,T')^F \cong N(T,T')^F/T'^F$, using Lang's Theorem, as in the proof of Lemma 2.6.

<u>Theorem 6.12</u> (Strong orthogonality) ([18], Theorem 6.2).

Let T,T' be two F-stable maximal tori in G. If (T,θ^{-1}), (T',θ') $(\theta \in \hat{T}^F, \theta' \in \hat{T}'^F)$ are not geometrically conjugate, then $(H_c^i(\tilde{X})_\theta \otimes H_c^j(\tilde{X}')_{\theta'})^{G^F} = 0$. [Here $\tilde{X},\tilde{X}'$ denote the schemes $L^{-1}U$, $L^{-1}U'$ with respect to some U,U' chosen such that TU, $T'U'$ are Borel subgroups.]

<u>Corollary 6.13</u> ([18], p. 136). If $(T,\theta),(T',\theta')$ are not geometrically conjugate then the virtual representations $R_T^G(\theta)$, $R_{T'}^G(\theta')$ are disjoint, i.e., have no irreducible constituents in common.

<u>Proof of Corollary</u>. This follows from the fact that $R_T^G(\theta^{-1})$ is the dual of $R_T^G(\theta)$; this follows, e.g., from the Character Formula (6.8).

<u>Proof of Theorem</u>. By the Künneth Formula (5.9) we have that

$$H_c^k(\tilde{X} \times \tilde{X})_{\theta,\theta'} \cong \sum_{i+j=k} H_c^i(X)_\theta \otimes H_c^j(\tilde{X})_{\theta'} ,$$

where $H_c^k(\tilde{X} \times \tilde{X})_{\theta,\theta'}$ is the subspace of $H_c^k(\tilde{X} \times \tilde{X}')$ on which $T^F \times T'^F$ acts on the right according to the character $\theta \times \theta'$. Thus we have to show that $(H^k(\tilde{X} \times \tilde{X}')_{\theta,\theta'})^{G^F} = 0$ if (T,θ^{-1}), (T',θ') are not geometrically conjugate. (Note that G^F acts

diagonally on the left on $\tilde{X} \times \tilde{X}'$ and hence on $H_c^i(\tilde{X} \times \tilde{X}')$.)
This means, by (5.10), that we have to show that
$(H_c^i(\tilde{X} \times \tilde{X}')/G^F)_{\theta,\theta'} = 0$.

Now we have an isomorphism

$$(\tilde{X} \times \tilde{X}')/G^F \to Y = \{(x,x',y) \in U \times U' \times G \mid x(Fy) = yx'\}$$

given by $(g,g') \to (g^{-1}(Fg), (g')^{-1}(Fg'), g^{-1}g')$,

and $T^F \times T'^F$ acts on Y as

$$(x,x',y) \to (t^{-1}\, x\, t,\ t'^{-1}yt',\ t^{-1}yt').$$

Let $Y_w = \{(x,x',y) \mid y \in G_w\}$; then Y_w is a locally closed subset of G_w. Now there is a filtration of G defined by the closures of the G_w (see e.g. A Borel and J. Tits, Publ. Math. IHES 41 (1972), §3) and this leads to a filtration of Y. Using the long exact sequence (5.8) repeatedly, we see that it is sufficient to show that $H_c^i(Y_w)_{\theta,\theta'} = 0$ for all w.

Now we embed $T^F \times T'^F$ in a torus H_w which acts on Y_w as follows. Let $H_w = \{(t,t') \in T \times T' \mid t'(Ft')^{-1} = (F\dot{w})^{-1}t(Ft)^{-1}(F\dot{w})$. Then $T^F \times T'^F \subset H_w \subset T \times T'$. Define an action of H_w on Y_w by the following rule. If $(t,t') \in H_w$, $f_{t,t'}(x,x',y) = (\tilde{x},\tilde{x}',\tilde{y})$, where

$$\tilde{x} = t^{-1}x(Fu_y)t(F(t^{-1}u_y{}^{-1}t))$$
$$\tilde{x}' = t'^{-1}x'(fu'_y)^{-1}t'F(t'^{-1}u'_y t')$$
$$\tilde{y} = t^{-1}yt'.$$

Then we can check that $(\tilde{x},\tilde{x}',\tilde{y}) \in Y_w$ and that $f_{t_1,t_1'} f_{t_2 t_2'} = f_{t_1 t_2, t_1' t_2'}$. Thus H_w acts on Y_w and then by Theorem 6.5

the action of H_w^0 on $H_c^i(Y_w)$ is trivial. Suppose that $H_c^i(Y_w)_{\theta,\theta'}$ is non-zero for some w. Then it follows that the restriction of the character $\theta\theta'$ of $T^F \times T'^F$ to $H_w^0 \cap (T^F \times T'^F)$ is trivial. From this we will deduce that (T,θ^{-1}) and (T',θ') are geometrically conjugate.

Now H_w is the kernel of the composite map $T \times T' \to T \times T' \to T$ given by

$$(t,t') \to ((Ft)t^{-1},(Ft')t'^{-1}) \to (F\dot{w})Ft')t'^{-1}(F\dot{w})^{-1}t(Ft)^{-1}.$$

Hence $Y(H_w^0)$, the group of one-parameter subgroups of H_w^0, is the kernel of the composite map

$$Y(T) \times Y(T') \to Y(T) \times Y(T') \to Y(T) \quad \text{given by}$$

$$(x,x') \to ((F-1)x,(F-1)x') \to (F\dot{w})(F-1)x' - (F-1)x.$$

[Note that the mapping $F\dot{w}$: $T' \to T$ given by $t' \to (F\dot{w})t'(F\dot{w})^{-1}$ induces a mapping Fw: $Y(T') \to Y(T)$.]

Now we make the following remark. Let S,S' be two tori of G with $S' \subset S$, and suppose S is F-stable. Then if $\theta \in \hat{S}^F$ is trivial on $S' \cap S^F$ then, regarded as a character of $Y(S)$ it is trivial on $((F-1)Y(S') \otimes \mathbb{Q}) \cap Y(S)$. This comes from looking at the inverse image of $S' \cap S^F$ under the map $Y(S) \to S^F$ given by (6.9).

Using this we see that $\theta\theta'$ is trivial on $((F-1)Y(H_w^0) \otimes \mathbb{Q}) \cap (Y(T) \times Y(T'))$. Now $F - 1$ is injective and $Y(T) \times Y(T')/(F - 1)(Y(T) \times Y(T'))$ is a torsion group. So the kernel M of the map $Y(T) \times Y(T') \to Y(T)$ given by $(x,x') \to (F\dot{w})x' - x$ must be contained in the above intersection. Hence $\theta\theta'$ is trivial on M and this says that for all x in $Y(T)$, $\theta(x)\theta'((F\dot{w})^{-1}(x)) = 1$. Thus $\theta\cdot{}^{(F\dot{w})}\theta' = 1$ as characters of $Y(T)$ and θ^{-1} and θ' are geometrically conjugate, as required.

Theorem 6.14 (Weak orthogonality) ([18], Theorem 6.8; [48], Theorem 2.3)

Let $\theta \in \hat{T}^F$, $\theta' \in \hat{T}'^F$. Then $(R_T^G(\theta), R_{T'}^G(\theta')) = \{w \in W(T,T')^F \mid {}^{\dot{w}}\theta' = \theta\}$. In particular, $(R_T^G(\theta), R_{T'}^G(\theta')) = 0$ (but $R_T^G(\theta)$, $R_{T'}^G(\theta')$ need not be disjoint) if T and T' are not G^F-conjugate.

Theorem 6.15 (Orthogonality relations for Green functions) ([18], Theorem 6.9)).

We have the following relations for the functions Q_T^G on the unipotent elements of G^F.

$$(6.16) \qquad \frac{1}{|G^F|}\sum_{\substack{u \in G^F \\ u \text{ unipotent}}} Q_T^G(u)Q_{T'}^G(u) = \frac{|N(T,T')^F|}{|T^F||T'^F|}$$

Proofs of Theorems 6.14 and 6.15. The proofs are by formal computation using the character formula (6.9). The strong orthogonality theorem (6.12) also enters into the proof at the end. We make an assumption (by induction on the semi-simple rank of G) that (6.16) holds for the groups $C^0(s)^F$, where s does not lie in $Z(G^F)$, the center of G^F. Then

$$(R_T^G(\theta),R_{T'}^G(\theta')) = \frac{1}{|G^F|}\sum_{h\in G^F} \mathrm{Tr}(h,R_T^G(\theta))\mathrm{Tr}(h^{-1},R_{T'}^G(\theta'))$$

$$= \frac{1}{|G^F|}\sum_{\substack{s\in G^F\\ s \text{ semisimple}}} \frac{1}{|C^0(s)|^2}\sum_{\substack{g,g'\in G^F\\ g^{-1}sg\in T^F\\ g'^{-1}sg\in T'^F}} \theta(g^{-1}sg)\overline{\theta'(g'^{-1}sg')}$$

$$\cdot \sum_{\substack{u\in C^0(s)^F\\ u \text{ unipotent}}} Q^{C^0(s)}_{gTg^{-1}}(u)\, Q^{C^0(s)}_{g'T'g'^{-1}}(u),$$

using (6.9),

$$= \frac{1}{|G^F|}\sum_{\substack{s\in G^F\\ s \text{ semisimple}\\ s\notin Z(G^F)}} \frac{1}{|C^0(s)^F|}\sum_{g,g'} \theta(g^{-1}sg)\ \overline{\theta'(g'^{-1}s'g')}$$

$$\cdot \frac{|N_{C^0(s)}(gTg^{-1},g'T'g'^{-1})^F|}{|T^F||T'^F|}$$

$$+ \sum_{\substack{s\in G^F\\ s\in Z(G^F)}} \frac{1}{|C^0(s)^F|}\Big(\frac{1}{|G^F|}\sum_{u \text{ unipotent}} Q_T^G(u)Q_{T'}^G(u)\Big)$$

$$= \frac{1}{|G^F|}\sum_{\substack{s\in G^F\\ s \text{ semisimple}}} \frac{1}{|C^0(s)^F|}\sum_{g,g'} \theta(g^{-1}sg)\ \overline{\theta'(g'^{-1}sg')}$$

$$\cdot \frac{|N_{C^0(s)}(gTg^{-1},g'T'g'^{-1})^F|}{|T^F||T'^F|} + A$$

where the sum is over all semisimple elements s including the central elements of G^F, and

$$A = \sum_{s\in Z(G^F)} \theta(s)\overline{\theta'(s)}\Big\{\frac{1}{|G^F|}\sum_{u \text{ unipotent}} Q_T^G(u)Q_{T'}^G(u) - \frac{N(T,T')^F}{|T^F||T'^F|}\Big\}.$$

Now the set

$$\{(g,g',n_1) \in G^F \times G^F \times G^F \mid g^{-1}sg \in T^F, g'^{-1}sg' \in T'^F,$$

$$n_1 \in N_{C^0(s)}(gTg^{-1}, g'T'g'^{-1})^F\}$$

is in bijection with the set

$\{(g,n,x) \in G^F \times N(T,T')^F \times C^0(s)^F \mid g^{-1}sg \in T^F\}$

under the map $(g,g',n_1) \leftrightarrow (g,n,n_1)$ where $n = g^{-1}n_1g'$, for $n \in N(T,T')^F$ is equivalent to $n_1 \in N_{C^0(s)}(gTg^{-1}, g'T'g'^{-1})^F$.

Hence

$$(R_T^G(\theta), R_{T'}^G(\theta')) = A + \frac{1}{|G^F|} \sum_{\substack{s \in G^F \\ s \text{ semisimple}}} \frac{1}{|C^0(s)^F|}$$

$$\sum_{\substack{g \in G^F \\ g^{-1}sg \in T^F \\ n \in N(T,T')^F}} \theta(g^{-1}sg)\, \overline{\theta'(ng^{-1}sgn^{-1})}\, \frac{|C^0(s)^F|}{|T^F||T'^F|}$$

$$= A + \frac{1}{|G^F|} \sum_{\substack{t \in T^F \\ n \in N(T,T')^F}} \frac{|G^F|}{|T^F||T'^F|}\, \theta(t)\, \overline{\theta'(ntn^{-1})}$$

$$= A + |\{w \in W(T,T')^F \mid {}^{\dot{w}}\theta = \theta'\}|,$$

since $\displaystyle \frac{1}{|T^F|} \sum_{t \in T^F} \theta(t)\overline{\theta'(ntn^{-1})} = \begin{cases} 0 & \text{if } \theta \neq {}^n\theta' \\ 1 & \text{if } \theta = {}^n\theta'. \end{cases}$

By (6.13) we have $(R_T^G(\theta), R_{T'}^G(1)) = 0$ if $\theta \neq 1$, since the only character of T^F geometrically conjugate to $1 \in \hat{T}'^F$ is $1 \in \hat{T}^F$. So if T^F has a non-trivial character θ which restricts to 1 on $Z(G^F)$ then we have $A = 0$, and $\sum_{s\varepsilon G^F} \theta(s)1(s^{-1})$ $\neq 0$, so that (6.16) holds. Then $A = 0$ in general and so Theorem 6.14 holds. Similarly our conclusions hold if T'^F has such a character. Now suppose neither T^F nor T'^F has such a character. Then $T^F = T'^F = Z(G^F)$. Then we compute $(R_T^G(1), R_T^G(1)) = (\mathrm{Ind}_{B_0^F}^{G^F}(1), \mathrm{Ind}_{B_0^F}^{G^F}(1))$ since T, T' must be split over $\mathbb{F}_q$. Then (6.16) holds by (3.2) and so $A = 0$ and Theorem 6.14 holds in this case also. This completes the proof.

Remarks.

1. The construction of the virtual representations $R_T^G(\theta)$ of G^F gives us a map $\theta \to R_T^G(\theta)$ from $\mathcal{R}(T^F)$ into $\mathcal{R}(G^F)$ where $\mathcal{R}(G^F)$ is the Grothendieck group of G^F. Let $\theta, \phi \; \varepsilon \; \hat{T}^F$. Using (3.1) and (6.14) we see that $(\mathrm{Ind}_{T^F}^{N(T)^F}(\theta), \mathrm{Ind}_{T^F}^{N(T)^F}(\phi))_{N(T)^F} = (R_T^G(\theta), R_T^G(\phi))_{G^F}$. Taking characters we get an isometry from the space of class functions on $N(T)^F$ vanishing outside T^F into the space of class functions on G^F. This is a familiar situation in the character theory of finite groups; see e.g. [26], p. 172. We also remark that if T and T' are not G^F-conjugate, so that $(R_T^G(\theta), R_{T'}^G(\phi)) = 0$ for $\theta \; \varepsilon \; \hat{T}^F$, $\phi \; \varepsilon \; \hat{T}'^F$, then in fact we have $\sum_g \mathrm{Tr}(g, R_T^G(\theta))\mathrm{Tr}(g^{-1}, R_{T'}^G(\phi)) = 0$, where the sum runs

over all $g = su$ where s is conjugate to a fixed semisimple element s_0. This follows from the character formula (6.9) and the orthogonality of Green functions (6.15). In other words, we have "section-by-section orthogonality" among the $R_T^G(\theta)$, where by a section we mean a p'-section.

2. Green [30] and Kilmoyer (unpublished) have defined the notion of the <u>principal part</u> at T of a class function f on G^F. This is defined by

$$f_{(T)}(t) = \frac{|T^F|}{|C^0(t)^F|} \sum_{\substack{u \in C(t)^F \\ u \text{ unipotent}}} f(tu) Q_T^{C^0(t)}(u) \qquad (t \in T^F).$$

The mapping $\theta \to \mathrm{Tr} R_T^G(\theta)$ $(\theta \in \hat{T}^F)$ extends by linearity to a map from the space of class functions on T^F into the space of class functions on G^F. If ϕ is a class function on T^F and the class function ϕ^* on G^F corresponds to ϕ under this map we have $(\phi^*, f)_{G^F} = (\phi, f_{(T)})_{T^F}$. Furthermore we have $f = \sum_{(T)} \frac{1}{|W(T)^F|}(f_{(T)})^*$ where the summation is over a set of representatives for the conjugacy classes of tori in G^F. The maps $\theta \to \mathrm{Tr}\, R_T^G(\theta)$ and $f \to f_{(T)}$ can be regarded as twisted versions of the induction and restriction maps for class functions.

<u>Definition</u>. The character $\theta \in \hat{T}^F$ is regular if it is not fixed by any non-trivial element of $W(T)$.

<u>Corollary 6.17</u>. If $\theta \in \hat{T}^F$ is regular then $\pm R_T^G(\theta)$ is irreducible.

Proof. This is clear from (6.14).

Corollary 6.18 ([48], 2.4) The virtual representation $R_T^G(\theta)$ is independent of the choice of U which defines the variety $\tilde{X} = L^{-1}U$.

Proof. Suppose f, f' are two class functions on G^F such that $(f,f) = (f,f') = (f',f') = 0$. Then $(f-f', f-f') = 0$ and hence $f = f'$. If we have $B = TU$, $B' = TU'$, we apply this remark to $f = R_{T,U}^G(\theta)$, $f' = R_{T,U'}^{G'}(\theta)$ and use Theorem 6.14.

Remark. We will show later (6.24) that every irreducible character of G^F occurs as a constituent of some $R_T^G(\theta)$. Corollary 6.17 and an asymptotic result on the number of regular characters of tori of G^F (see [72], pp. 638, 640) show that the $R_T^G(\theta)$, for θ regular, give "most" of the irreducible characters of G^F, in some sense.

Next we determine the dimensions of the $R_T^G(\theta)$. The ubiquitous Steinberg representation enters into this calculation. We first define this representation and review some of its properties. Any F-stable maximal torus T of G has a decomposition $T = T_d T_a$, where T_d is a maximal F_q-split subtorus of T (see e.g. [5], p. 218). Let $\sigma(G) = \dim(T_0)_d$. We define $\varepsilon_G = (-1)^{\sigma(G)}$.

The Tits building of G^F is the simplicial complex $\mathcal{J}$ whose simplices are indexed by the parabolic subgroups of G^F. We denote the simplex corresponding to P^F by S_P. We say S_P is a face of S_Q if $Q \subset P$. Then G^F acts on $\mathcal{J}$ (by conjugation) and hence on the (classical) cohomology groups

$H^i(\mathcal{J},\mathbb{Z})$. The representation of G^F on the top cohomology group is called the Steinberg representation and denoted by St_G. It has dimension equal to $|U_0^F|$ and this number is also equal to $|G^F|_p$, the order of a Sylow p-subgroup of G^F. We state some properties of St_G that we will need (see [73], §5; [14], 4.2.8).

(6.19) $\quad (\mathrm{Ind}_{B_0^F}^{G^F}(1), St_G) = 1.$

(6.20) The character of St_G is given as follows. If g is not semisimple, $St_G(g) = 0$. If s is semisimple, $St_G(s) = \varepsilon_G \varepsilon_{C^0(s)} St_{C^0(s)}(1)$.

<u>Theorem 6.21</u> ([18], 7.1; [48], 2.9).

$$\begin{aligned}\dim R_T^G(\theta) &= Q_T^G(1)\\ &= \varepsilon_G\varepsilon_T \frac{|G^F|}{|U_0^F|\,|T^F|}\\ &= \frac{\varepsilon_G\varepsilon_T|G^F|}{St_G(1)\,|T^F|}\end{aligned}$$

[The content of the theorem is that although there is no decomposition $B = TU$ in general with an F-stable B, the dimension of $R_T^G(\theta)$ is the same as though we were inducing from such a subgroup B^F.]

<u>Proof</u>. First assume that we have a non-trivial $\theta \in \hat{\mathbf{T}}^F$ such that θ is trivial on $Z(G^F)$. Then by (6.13) $R_T^G(\theta)$ and $R_{T_0}^G(1) = \mathrm{Ind}_{B_0^F}^{G^F}(1)$ are disjoint and thus $(R_T^G(\theta), St_G) = 0$ by (6.19). Using (6.8) and (6.20) we get

$$\sum_{\substack{s \in G^F \\ s \text{ semisimple}}} \varepsilon_G \varepsilon_{C^0(s)} \frac{St_{C^0(s)}(1)}{|C^0(s)^F|} \sum_{\substack{g \in G^F \\ g^{-1}sg \in T}} Q^{C^0(s)}_{gTg^{-1}}(1)\theta(g^{-1}sg) = 0.$$

We make the induction assumption that (6.21) is true for the groups $C^0(s)$ if s is a non-central semisimple element of G^F. This means that for such elements s we can put

$Q^{C^0(s)}_{gTg^{-1}}(1) = \varepsilon_{C^0(s)} \varepsilon_T \dfrac{|C^0(s)^F|}{St_{C^0(s)}(1)\,|T^F|}$. So we get

$$\varepsilon_G \varepsilon_T \frac{1}{|T^F|} \sum_{\substack{s \in G^F \\ s \text{ semisimple} \\ s \notin Z(G^F)}} \sum_{\substack{g \in G^F \\ g^{-1}sg \in T^F}} \theta(g^{-1}sg)$$

$$+ \sum_{\substack{s \in G^F \\ s \in Z(G^F)}} \frac{St_G(1)}{|G^F|} \sum_{\substack{g \in G^F \\ g^{-1}sg \in T}} Q^G_T(1)\theta(s) = 0,$$

i.e., $\varepsilon_G \varepsilon_T \dfrac{|G^F|}{|T^F|} \sum\limits_{\substack{t \in T^F \\ t \notin Z(G^F)}} \theta(t) + \sum\limits_{z \in Z(G^F)} St_G(1) Q^G_T(1)\theta(z) = 0.$

Since $\sum\limits_{t \notin Z(G^F)} \theta(t) = - \sum\limits_{z \in Z(G^F)} \theta(z) \neq 0$, we have

$St_G(1) Q^G_T(1) = \varepsilon_G \varepsilon_T \dfrac{|G^F|}{|T^F|}$, as required.

Now assume there is no $\theta \in \hat{T}^F$ with the required properties. Then $T^F = Z(G^F)$ and we must have $R^G_T(\theta) \cong \mathrm{Ind}^{G^F}_{B_0^F}(1)$ and $Q^G_T(1) = \dfrac{|G^F|}{|B_0^F|}$, again agreeing with the statement of the theorem.

<u>Corollary 6.22</u> ([18], 7.3). $\operatorname{Ind}_{T^F}^{G^F}(\theta) = \varepsilon_G\varepsilon_T\, R_T^G(\theta) \otimes St_G$.

<u>Proof</u>. By (6.20) it is sufficient to check that the traces of both sides are equal at semisimple elements of G^F. Let $s \in G^F$ be semisimple. By (6.8) we have

$$\operatorname{Tr}(s,R_T^G(\theta)) = \frac{1}{|C^0(s)^F|} \sum_{\substack{g\in G^F \\ g^{-1}sg\in T^F}} Q_{gTg^{-1}}^{C^0(s)}(1)\theta(g^{-1}sg)$$

$$= \frac{\varepsilon_{C^0(s)}\varepsilon_T}{St_{C^0(s)}(1)\,|T^F|} \sum_{\substack{g\in G^F \\ g^{-1}sg\in T^F}} \theta(g^{-1}sg).$$

Thus $\operatorname{Tr}(s,\varepsilon_G\varepsilon_T R_T^G(\theta) \otimes St_G) = \frac{1}{|T^F|} \sum_{\substack{g\in G^F \\ g^{-1}sg\in T^F}} \theta(g^{-1}sg)$

$$= \operatorname{Tr}(s,\operatorname{Ind}_{T^F}^{G^F}(\theta)).$$

The next theorem shows that every irreducible representation of G^F occurs as a constituent of some $R_T^G(\theta)$.

<u>Theorem 6.23</u> ([18], 2.11). The regular representation of G^F is equal to $\frac{1}{|G^F|_p} \sum_T \sum_{\theta\in\hat{T}} \varepsilon_T\varepsilon_G R_T^G(\theta)$, where the first summation is over all F-stable maximal tori T of G.

<u>Proof</u>. Denote the above expression, which is an element of $\mathcal{R}(G^F) \otimes_{\mathbb{Z}} \mathbb{Q}$, by ϕ. We compute (ϕ,ϕ) using (6.14). We have

$$(\phi,\phi) = \frac{1}{|G^F|_p^2} \sum_{T,T'} \sum_{\substack{\theta\in\hat{T}^F \\ \theta'\in\hat{T}'^F}} \varepsilon_T\varepsilon_{T'} |\{n \in N(T,T')^F | {}^n\theta = \theta'\}| \frac{1}{|T^F|}$$

$$= \frac{1}{|G^F|_p^2} \sum_{\substack{T,T' \\ \theta,\theta'}} \varepsilon_T\varepsilon_{T'} |\{g \in G^F | gTg^{-1} = T' \text{ and } {}^g\theta = \theta'\}| \frac{1}{|T^F|}$$

$$= \frac{1}{|G^F|_p^2} \sum_{\substack{T,\theta \\ g\in G^F}} \varepsilon_T\varepsilon_{gTg^{-1}} \frac{1}{|T^F|}$$

$$= \frac{1}{|G^F|_p^2} \cdot |G^F||G^F|_p^2$$

$$= |G^F|,$$

using the fact that the number of F-stable maximal tori of G is $|G^F|_p^2$ (see 5.13).

Let ρ denote the regular representation of G^F. Since $(\rho, R_T^G(\theta)) = \dim R_T^G(\theta)$ we have

$$(\rho,\phi) = \frac{1}{|G^F|_p} \sum_{T,\theta} \varepsilon_T\varepsilon_G\varepsilon_G\varepsilon_T \frac{|G^F|_{p'}}{|T^F|}$$

$$= |G^F|.$$

Since we also have $(\rho,\rho) = |G^F|$, we must have $(\rho-\phi,\rho-\phi) = 0$ and hence $\rho = \phi$.

Remark. The characters of the $R_T^G(\theta)$ do not span the space of class functions on G^F, in general; they do in the case of GL_n, but not even in the case of SL_2. The subspace of the space of class functions which is spanned by the characters of the $R_T^G(\theta)$ is called the space of uniform functions on G^F.

We now consider the connections between the $R_T^G(\theta)$ and the Harish-Chandra theory described in Chapter IV.

Theorem 6.24 ([18], 8.2; [48], 2.6). Suppose the F-stable maximal torus T is contained in an F-stable parabolic subgroup P of G. Let $P = LV$ where L is an F-stable Levi subgroup. Let $\theta \in \hat{T}^F$. Then $R_T^G(\theta) = \mathrm{Ind}_{P^F}^{G^F}(\widetilde{R_T^L(\theta)})$. [Here, as in Chapter IV, if M is an L^F-module, $\tilde{M}$ denotes its lift to P^F.]

Proof. Choose a Borel subgroup $B = TU$ where $T \subset B \subset P$, and B is not necessarily F-stable. As before, let $\tilde{X} = L^{-1}U$. Then if $g \in \tilde{X}$, $Fg = gu$ for some $u \in U \subset P$. Then $F(gPg^{-1}) = gPg^{-1}$ and so gPg^{-1} is F-stable.

Let $\mathcal{S}$ be the (finite) set of all F-stable conjugates of P. This set is the same as the set of G^F-conjugates of P, by (2.7), since $N_G(P) = P$. We have a map $\phi: g \to gPg^{-1}$ of $\tilde{X}$ onto $\mathcal{S}$. Let $P' \in \mathcal{S}$, and suppose $P' = g_1Pg_1^{-1}$, where $g_1 \in G^F$. Then the fibre $\tilde{X}_{P'}$ of ϕ over P' is the set $\{g \in \tilde{X} \mid gPg^{-1} = g_1Pg_1^{-1}\} = \{g_1^{-1}g \in P \mid (g_1^{-1}g)^{-1}F(g_1^{-1}g) \in U\}$. Now we have a natural map $\tilde{X}_{P'} \to \bar{\tilde{X}}_{P'} = \{\overline{g_1^{-1}g} \in L \mid (\overline{g_1^{-1}g})^{-1}\overline{F(g_1^{-1}g)} \in \bar{U}\}$ where for any set $Y \subset P$, $\bar{Y}$ denotes the image of Y under the canonical map $P \to L$. Furthermore, $\bar{\tilde{X}}_{P'}$ is of the form $L^{-1}\bar{U}$ for L. The fibres of this map are all isomorphic to V, which is isomorphic to affine space of dimension s, say. Then by (5.5) and (5.7) we get $H_c^i(\tilde{X}_{P'}) \cong H_c^{i-2s}(\bar{\tilde{X}}_{P'})$. [We disregard the Tate twist here; it is of significance only when calculating the action of Frobenius on cohomology.]

Now we have $\tilde{X} = \bigcup_{P' \in \mathcal{S}} \tilde{X}_{P'}$. We look at the action of G^F on $\tilde{X}$. Since G^F acts transitively on $\mathcal{S}$ by conjugation it permutes the schemes $\tilde{X}_{P'}$. Thus G^F permutes the cohomology groups $H_c^i(\tilde{X}_{P'})$ (for a fixed i) transitively, and the stabilizer of one of them, i.e., of P, is P^F. Now we make the following simple observation: Let $Q < H$ be finite groups and let M be an H-module. Let N be a Q-submodule of M such that $M = x_1N \oplus x_2N \oplus \cdots \oplus x_rN$ where $\{x_i\}$ is a set of left coset representatives for H over Q. Then $M = \mathrm{Ind}_Q^H(N)$. This fact has an obvious extension to virtual representations. We also have the additivity formula (6.1) for the alternating trace of G^F on $\oplus H_c^i(\tilde{X})$ in terms of the alternating traces on the $\oplus H_c^i(\tilde{X}_{P'})$. This proves the theorem.

<u>Definition</u>. An F-stable maximal torus T is said to be <u>minisotropic</u> if T is not contained in any F-stable proper parabolic subgroup of G.

<u>Example</u>. If $G = GL_n$, F the standard Frobenius map, the maximal torus T which corresponds to the "Coxeter element" of $W(T_0)$ (i.e., to an n-cycle in S_n) is a minisotropic torus and is the only one up to conjugacy by G^F.

<u>Theorem 6.25</u> ([18], 8.3; [48], 2.18). Let $\mu \in (G)$. Then μ is cuspidal if and only if $(\mu, R_T^G(\theta))_{G^F} = 0$ whenever T is not minisotropic, for any $\theta \in \hat{T}^F$.

<u>Proof</u>. Suppose $(\mu, R_T^G(\theta))_{G^F} = 0$ for any F-stable maximal torus T which is contained in L, where $P = LV$ is an F-stable

proper parabolic subgroup of G and L is an F-stable Levi subgroup of P. If ρ_L denotes the regular representation of L^F we have $\mathrm{Ind}_{V^F}^{G^F}(1) = \mathrm{Ind}_{P^F}^{G^F}(\tilde{\rho}_L)$

$$= \mathrm{Ind}_{P^F}^{G^F}\{\frac{1}{|L^F|_p} \sum_{\substack{T \subset L \\ \theta \in \hat{T}^F}} \varepsilon_L \varepsilon_T R_T^L(\theta)\} \quad \text{by (6.23)}$$

$$= \frac{1}{|L^F|_p} \sum_{\substack{T \subset L \\ \theta \in \hat{T}^F}} \varepsilon_L \varepsilon_T R_T^G(\theta) \quad \text{by (6.24)}$$

Now we have $(\mu, \mathrm{Ind}_{V^F}^{G^F}(1)) = 0$ and this shows that μ is cuspidal.

Conversely, suppose $(\mu, R_T^G(\theta)) \neq 0$ for some $T \subset L$, $P = LV$, $\theta \in \hat{T}^F$. Then $(\mu, R_T^G(\theta))_{G^F} = (\mu, \mathrm{Ind}_{P^F}^{G^F}(\widetilde{R_T^L(\theta)}))_{G^F} = (\mu, R_T^L(\theta))_{P^F} \neq 0$.

Let $\mu = \mu' + \mu''$ where μ' is the representation of P^F on the space of V^F-fixed vectors in the representation space of μ. Since V^F acts trivially on $R_T^L(\theta)$, $(\mu'', \widetilde{R_T^L(\theta)})_{P^F} = 0$. So $(\mu', \widetilde{R_T^L(\theta)})_{P^F} \neq 0$. Hence there exist non-zero V^F-fixed vectors in μ and μ is not cuspidal. This proves the theorem.

In the light of Theorems 6.24 and 6.25 we make the following remarks. For any F-stable proper parabolic subgroup $P = LV$ where L is F-stable, the cuspidal representations of L^F occur as constituents of some $R_T^L(\theta)$ where T is minisotropic in L. We can lift these representations of L^F to P^F and induce them to G^F. The irreducible representations of G^F which do not arise in this way are the cuspidal representations of G^F, which are constituents of the $R_T^G(\theta)$ where T is minisotropic in G. Of particular interest are the cuspidal representations which are constituents of some $R_T^G(1)$. These will be studied in more detail in Chapter VIII.

CHAPTER VII. CHARACTERS.

From the Character Formula (6.9) we see that in order to be able to write down the characters of the $R_T^G(\theta)$ we have to know the Green functions $Q_T^G(u)$. So far we only know that $Q_T^G(u)$ is an integer. Except in the case of some groups of low rank where character tables are known ([21], [22], [23], [57], [10], [76], [87]), in the case of GL_n (see [30], [52]) where there is a recursive formula to determine the functions, and in the case where T is a "Coxeter torus" in a classical group (see [46]), no explicit formulae are known for the Green functions in general. In this chapter we describe the work of Springer [75] and Kazhdan [41] which enable us to write down, for sufficiently large p, expressions for the $Q_T^G(u)$ as "trigonometric sums" on the Lie algebra of the group.

Consider the case when $T = T_0$. In this case we have a geometric interpretation of the $Q_{T_0}^G(u)$, since $R_{T_0}^G(1)$ is the permutation representation of G^F on the cosets of B_0^F.

Definition. If u is a unipotent element of G, $\mathcal{B}_u$ is the variety of all Borel subgroups of G containing u. We denote $\mathcal{B}_1$ by $\mathcal{B}$.

We note that $\mathcal{B}_u$ is a projective variety since it is a subvariety of the variety $\mathcal{B}$ of all Borel subgroups of G. (See [80], where this variety is studied.) For example, if $G = GL_n$, $\mathcal{B}_u$ is the variety of all flags (in the vector space

V on which GL_n acts) fixed by u. Now if u is fixed by F we have an action of F on $\mathcal{B}_u$ and hence on $H_c^i(\mathcal{B}_u)$. We then see that $Q_{T_0}^G(u) = |\mathcal{B}_u^F| = \sum_i (-1)^i \operatorname{Tr}(F, H_c^i(\mathcal{B}_u))$ by (5.11). This is a cohomological interpretation for $Q_{T_0}^G(u)$, and we would like to have a similar interpretation for $Q_T^G(u)$ for any T. This is done by defining an action of $W = W(T_0)$ on $H_c^i(\mathcal{B}_u)$ (although W does not act on $\mathcal{B}_u$ itself) and then, if T corresponds to $w \in W$ (in the sense of (2.8)), by obtaining $Q_T^G(u)$ as the alternating trace of "F twisted by w" on $\oplus H_c^i(\mathcal{B}_u)$. There are three steps in this process, the first two being due to Springer ([74], [75]) and the third, for sufficiently large p, due to Kazhdan [41]. They are as follows. We assume that we have a fixed F-stable maximal torus T. As before, $K = \bar{\mathbb{F}}_q$.

(i) Let A be an F-stable nilpotent element in the Lie algebra $\underline{\underline{g}}$ of G. We write down a trigonometric function" on $\underline{\underline{g}}$ corresponding to A.

(ii) Express this trigonometric sum as the alternating trace of an operator on the cohomology of $\mathcal{B}_A$, where $\mathcal{B}_A$ is the variety of all Borel subgroups of G whose Lie algebras contain A.

(iii) Lift the function so obtained on the F-stable nilpotent elements of $\underline{\underline{g}}$ to a function on the F-stable unipotent elements of G, by using an exponential map. Prove that the function thus obtained on the unipotent elements of G^F is (up to a constant) the Green function Q_T^G.

We start with some preliminaries on the Lie algebra of G (see eg. [5], p. 117 or [40], p. 65).

Definition. The Lie algebra $\underline{g}$ of G is the K-space of all left-invariant derivations of $K[G]$ made into a Lie algebra by the rule $[\delta_1,\delta_2] = \delta_1\delta_2 - \delta_2\delta_1$. In other words, $\underline{g} = \{\delta | \delta$ is a derivation of $K[G]$ and $\delta\lambda_x = \lambda_x\delta$ for all $x \in G\}$.

The group G acts on $\underline{g}$ by the Adjoint representation which can be described as $\mathrm{Ad}\, g\colon \delta \to \rho_g \delta \rho_g^{-1}$ ([40], p. 66). For example, if $G = GL_n$, $\underline{g}$ is the Lie algebra of all $n \times n$ matrices over K and for $g \in G$, $\mathrm{Ad}\, g$ is the map $X \to gXg^{-1}$.

Let $B = TU$ be a Borel subgroup of G. Then we have a decomposition $\underline{g} = \underline{t} + \underline{n} + \underline{n}^-$, where $\underline{t},\underline{n},\underline{n}^-$ are the Lie algebras of T, U, U^- respectively. We have $\underline{n} = \sum_{\substack{\alpha\in\Phi \\ \alpha>0}} KX_\alpha$, $\underline{n}^- = \sum_{\substack{\alpha\in\Phi \\ \alpha<0}} KX_\alpha$, where the X_α are "root vectors" and KX_α is the Lie algebra of the root subgroup U_α in G.

Let $\underline{g}'$ be the vector space dual of $\underline{g}$ and $<,>$ the pairing between $\underline{g}$ and $\underline{g}'$. If G is an algebraic group over $\mathbb{C}$, $\underline{g}$ and $\underline{g}'$ can be identified by the Killing form, which is a non-degenerate, symmetric bilinear form on $\underline{g}$. For large p we have this also in characteristic p. For example, if

$G = GL_n$, the form given by $\langle X,Y\rangle = \mathrm{Tr}\, XY$ has the above properties.

Definition. If A is a nilpotent element in $\underline{g}$, $\mathcal{B}_A$ is the variety of all Borel subgroups of G whose Lie algebras contain A.

We observe that there is a natural action of F on $\underline{g}$, and we denote the set of F-fixed points by $\underline{g}^F$. We have morphisms $\mathrm{Ad}:G \to \mathrm{End}\,\underline{g}$, $\mathrm{Ad}':G \to \mathrm{End}\,\underline{g}'$, and it can be shown, using their definitions, that they are defined over $\mathbb{F}_q$. Let $\underline{t}'$ be the subspace of $\underline{g}'$ which is orthogonal to $\underline{n} + \underline{n}^-$. Then $\underline{t}'$ is isomorphic to the dual of $\underline{t}$.

Definition. $A' \in \underline{t}'$ is strongly regular if $C_G(A') = T$.

We now fix a non-trivial additive character $\psi:\mathbb{F}_q \to \bar{\mathbb{Q}}_\ell^*$.

Definition. If A is a nilpotent element of $\underline{g}^F$ and A' is a strongly regular element of $\underline{t}'^F$, then

$$S(A,A') = \sum_{X \in G^F\text{-orbit of } A'} \psi\langle A,X\rangle$$

Remarks Here the action of G^F is by Ad'. Note also that since $A \in \underline{g}^F$, $X \in \underline{g}'^F$, we have $\langle A,X\rangle \in \mathbb{F}_q$. Also this "trigonometric sum" is just $\sum_{g \in G^F/C_{G^F}(A')} \psi\langle A,(\mathrm{Ad}'g)A'\rangle$

$= \sum_g \psi\langle(\mathrm{Ad}\, g^{-1})A,A'\rangle$, which bears a resemblance to the value of an induced character, since $Y \to \langle Y,A'\rangle$ is a linear functional on $\underline{b} = \underline{n} + \underline{t}$.

<u>Definition</u>. If $a \in K$, $Y_{A,A',a} = \{(x,X) \in K\times(\mathrm{Ad}'G)A' \mid x^q - x = \langle A,X\rangle + a\}$.

Write $Y_{A,A'}$ for $Y_{A,A',0}$. Then F acts on $Y_{A,A'}$ by $(x,X) \to (x^q,FX)$ and the additive group $\mathbb{F}_q^+$ acts on $Y_{A,A'}$ by $\zeta(a):(x,X) \to (x+a,X)$ $(a \in \mathbb{F}_q^+)$. Let $\zeta(a)^i$ be the induced linear map on $H_c^i(Y_{A,A'})$. If $b \in K$ is such that $b^q - b = a$, the map $f: Y_{A,A'} \to Y_{A,A',a}$ given by $(x,X) \to (x+b,X)$ is an isomorphism and $Ff = f\zeta(a)F$. Thus we have that

$$\mathrm{Tr}(F,\ H_c^i(Y_{A,A',a})) = \mathrm{Tr}(F\ \zeta(a)^i,\ H_c^i(Y_{A,A'})).$$

We note that $(x,X) \in Y^F_{A,A',a}$ if and only if $x \in \mathbb{F}_q$ and $X \in (\mathrm{Ad}'G^F)A'$ with $\langle A,X\rangle = -a$. Hence we have

$$S(A,A') = \sum_{X\in(\mathrm{Ad}'G^F)A'} \psi\langle A,X\rangle = q^{-1} \sum_{a\in\mathbb{F}_q} \psi(-a)\,|Y^F_{A,A',a}|$$

$$= q^{-1} \sum_i (-1)^i[\psi(-a)\mathrm{Tr}(F\zeta(a)^i,\ H_c^i(Y_{A,A'}))], \text{ using (5.11)},$$

and so

$$(7.1)\quad S(A,A') = \sum_i (-1)^i\ \mathrm{Tr}(F,\ H_c^i(Y_{A,A'})_\psi).$$

Here $H_c^i(Y_{A,A'})_\psi$ is the subspace of $H_c^i(Y_{A,A'})$ consisting of all $x \in \mathbb{F}_q$ such that $\zeta(a)^i x = \psi(a)x$. Note that this is an F-stable subspace. We have also used the fact that if ϕ is any character of $\mathbb{F}_q^+$, then $q^{-1} \sum_{a\in\mathbb{F}_q} \psi(-a)\phi(a) = \begin{cases} 0, & \phi \neq \psi \\ 1, & \phi = \psi \end{cases}$.

In general these sums are not easy to compute explicitly. We give an example where we compute them.

Example. $G = SL_2$, $\underline{g} = s\ell_2$, $p \neq 2$, $A = \begin{pmatrix} 0 & 1 \\ 0 & 0 \end{pmatrix}$, $A' = \begin{pmatrix} 1 & 0 \\ 0 & -1 \end{pmatrix}$.

We identify $\underline{g}$ and $\underline{g}'$ by means of the form $\langle X.Y\rangle = \operatorname{Tr} XY$ on $\underline{g}$. Now, by the Bruhat decomposition, $G = B_0 \cup B_0 w B_0$ where $w = \begin{pmatrix} 0 & 1 \\ -1 & 0 \end{pmatrix}$. Also T_0 centralizes A'. If $g \in B_0^F$ then $\langle A, (\mathrm{Ad}g)A'\rangle = 0$. If $g \in (B_0wB_0)^F$, let $g = \begin{pmatrix} 1 & a \\ 0 & 1 \end{pmatrix}\begin{pmatrix} 0 & 1 \\ -1 & 0 \end{pmatrix}\begin{pmatrix} 1 & b' \\ 0 & 1 \end{pmatrix}$.

Then $\langle A,(\mathrm{Ad}g)A'\rangle = 2b$. If we choose $\psi:\mathbb{F}_q^+ \to \mathbb{C}^*$ to be a non-trivial character then we get $S(A,A') = q + \sum_{b\in F_q} \psi(2b) = q$.

Now suppose we take $A = 0$. Then we get by similar calculations that $S(A,A') = q + q^2$. So we get

$$q^{-1}S(A,A') = \begin{cases} 1, & A = \begin{pmatrix} 0 & 1 \\ 0 & 0 \end{pmatrix} \\ q+1, & A = 0. \end{cases}$$

These are precisely the values $Q^G_{T_0}(u)$ for $u = \begin{pmatrix} 1 & 1 \\ 0 & 1 \end{pmatrix}$ and $u = \begin{pmatrix} 1 & 0 \\ 0 & 1 \end{pmatrix}$ respectively (see the character table for GL_2 in Chapter III). If we were to choose A' to be an element whose centralizer is a non-split torus in G, we would end up with the values of the discrete series characters at the unipotent elements.

The next few theorems are aimed at relating the groups

$H^i_c(Y_{A,A'})_\psi$ with the cohomology of $\mathcal{B}_A$, and in the process we will define a representation of $W = W(T)$ on $H^i_c(\mathcal{B}_A)$. Recall that T is a fixed F-stable maximal torus, and A is a nilpotent element of $\underline{g}^F$.

<u>Definition</u>. $\underline{t}'_0$ is the set of strongly regular elements in $\underline{t}'$.

<u>Remark</u>. It is known that $\underline{t}'_0 \neq \phi$ if $p \neq 2$ and $\underline{t}_0'^F \neq \phi$ if q is sufficiently large. We assume from now on that $t_0'^F \neq \phi$.

<u>Definition</u>. $\mathcal{Y} = \{(x,gT,A') \in K \times G/T \times \underline{t}'_0 | x^q - x = \langle A,(Ad'g)A'\rangle\}$.

We have actions of $\mathbb{F}_q$, $W = W(T)$ and F on $\mathcal{Y}$:

$a \in \mathbb{F}_q$ acts by $(x,gT,A') \to (x+a,gT,A')$.

$w \in W$ acts by $(x,gT,A') \to (x,gw^{-1}T, (Ad\dot{w})A')$

F acts by $(x,gT,A') \to (x^q,(Fg)T, FA')$.

Here, as before, $\dot{w}$ is a representative in $N(T)$ for W. The actions of $\mathbb{F}_q$ and F commute, whereas $F \cdot w = (Fw) \cdot F$ $(w \in W)$.

We now choose a Borel subgroup $B = TU$ containing T, and note that B is not necessarily F-stable. Having made this choice of B in $\mathcal{B}$, the variety of Borel subgroups of G, we have maps

$$\mathcal{Y} \xrightarrow{f} \mathcal{B} \times \underline{t}'_0 \xrightarrow{\sigma} \underline{t}'_0 \text{ given by}$$

$$(x,gT,A') \xrightarrow{f} (gBg^{-1}, A') \xrightarrow{\sigma} A'.$$

Let $\pi = \sigma f: \mathcal{Y} \to \underline{t}'_0$. The action of W on $\mathcal{Y}$ and $\underline{t}'_0$ commutes with π. The idea then is to take the constant sheaf $\bar{\mathbb{Q}}_\ell$ on $\mathcal{Y}$, look at the direct image sheaf $R^i\pi_!\bar{\mathbb{Q}}_\ell$ on $\underline{t}'_0$, take a suitable constant subsheaf of it which has stalks isomorphic to the groups $H_c^{i-2d}(\mathcal{B}_A)$ (d=dim U) and get an action of W on these groups.

<u>Theorem</u> 7.2 ([75], 3.5, 4.1) (i) The direct image sheaf $R^i f_! \bar{\mathbb{Q}}_\ell = 0$ if $i \neq 2d$, where $d = \dim \mathcal{B} (= \dim U)$.

(ii) The sheaf $(R^{2d} f_! \bar{\mathbb{Q}}_\ell)_\psi$ on $\mathcal{B} \times \underline{\underline{t}}'_0$ is supported on $\mathcal{B}_A \times \underline{\underline{t}}'_0$ and its restriction to $\mathcal{B}_A \times \underline{\underline{t}}'_0$ is the constant sheaf $\bar{\mathbb{Q}}_\ell(-d)$

[Here $(R^{2d} f_! \bar{\mathbb{Q}}_\ell)_\psi$ denotes the part of $R^{2d} f_! \bar{\mathbb{Q}}_\ell$ on which $\mathbb{F}_q$ acts according to the character ψ. We let $\mathbb{F}_q$ act trivially on $\mathcal{B} \times \underline{\underline{t}}'_0$, so that f commutes with the actions of $\mathbb{F}_q$ on $\mathcal{Y}$ and $\mathcal{B} \times \underline{\underline{t}}'_0$. Thus we get an action of $\mathbb{F}_q$ on $R^{2d} f_! \bar{\mathbb{Q}}_\ell$.]

<u>Proof</u>. Consider a geometric point $\bar{s}$ of $\mathcal{B} \times \underline{\underline{t}}'_0$ centered at $s = (gBg^{-1}, A')$. By (5.3) the stalk of $R^i f_! \bar{\mathbb{Q}}_\ell$ at $\bar{s}$ is $H_c^i(\mathcal{Y}_{\bar{s}})$ where $\mathcal{Y}_{\bar{s}}$ is the fibre of f over $\bar{s}$. In our case s is a closed point and we are working over an algebraically closed field K. Hence the stalk of $R^i f_! \bar{\mathbb{Q}}_\ell$ at $\bar{s}$ is isomorphic to $H_c^i(f^{-1}s)$ where $f^{-1}s$ the usual fibre over s (see the remark in the section on geometric points, Chapter V and Remark 6.6). Now $f^{-1}s = \{x, hT, A') \mid hBh^{-1} = gBg^{-1}\}$. If $hBh^{-1} = gBg^{-1}$ then $g^{-1}h \in B$ and so $g^{-1}h = ut$ where $t \in T$, $u \in U$. Thus $f^{-1}s = \{(x, guT, A') \mid x^q - x = \langle (\mathrm{Ad}\, g^{-1})A, (\mathrm{Ad}' u)A' \rangle\}$.

Consider $(\mathrm{Ad}'U)A'$. We have $\underline{\underline{g}} = \underline{\underline{t}} \oplus \sum_{\alpha>0} KX_\alpha \oplus \sum_{\alpha<0} KX_\alpha$

where $\underline{t} \oplus \sum_{\alpha>0} KX_\alpha = \underline{b}$, the Lie algebra of B. Then we can write $\underline{g}' = \underline{t}' \oplus \sum_{\alpha>0} KX'_\alpha \oplus \sum_{\alpha<0} KX'_\alpha$ where $<\underline{t},X'_\alpha> = 0$, $<X_\alpha,X'_\beta> = \delta_{-\alpha,\beta}$. Using the formulas for the adjoint representation $\underline{g} \to \mathrm{End}\,\underline{g}$ (see eg. [39], p. 96) we can compute the action of the elements $x_\alpha(t) \in U_\alpha \subset U$ in the representations Ad, Ad' of G. We can see that if $u \in U$, $(\mathrm{Ad}'u)\,A' \in A' + \underline{b}^\perp$, where $\underline{b}^\perp = \sum_{\alpha>0} KX'_\alpha$ is the subspace of $\underline{g}'$ orthogonal to $\underline{b}$. By a theorem of Rosenlicht (see eg. [80], p. 35) any orbit of a unipotent group acting on an affine variety is closed and hence $(\mathrm{Ad}'U)A'$ is a closed subvariety of $\underline{g}'$. Since A' is strongly regular, it has the same dimension d as U, and d is also equal to $\dim \underline{b}^\perp$. Hence $(\mathrm{Ad}'U)A' = A' + \underline{b}^\perp$.

Now suppose $s \notin \mathcal{B}_A \times \underline{t}'_0$, ie. $gBg^{-1} \notin \mathcal{B}_A$. This means that $(\mathrm{Ad}g^{-1})A \notin \underline{b}$. Consider $<(\mathrm{Ad}g^{-1})A, \mathrm{Ad}'u)A'>$. Since $(\mathrm{Ad}'u)A'$ covers all the elements of $A' + \underline{b}^\perp$ as u runs over U, and U is isomorphic to affine space $\mathbb{A}^d$, we have that $f^{-1}s$ is isomorphic to a finite covering of $\mathbb{A}^d$. The action of $\mathbb{F}_q$ on the first coordinate of $f^{-1}s$ by translation can be extended to an action of K, and then we see by Theorem 6.5 that K, and hence $\mathbb{F}_q$, acts trivially on $H^i_c(f^{-1}s)$. Thus the group $(H^i_c(f^{-1}s))_\psi$, which is the part of $H^i_c(f^{-1}s)$ on which $\mathbb{F}_q$ acts according to the non-trivial character ψ, is zero for all i. If $(\mathrm{Ad}g^{-1})A \in \underline{b}$, then $<(\mathrm{Ad}g^{-1})A, (\mathrm{Ad}'u)A'> = 0$. In this case $f^{-1}s \simeq \mathbb{F}_q \times \mathbb{A}^d$, and $\mathbb{F}_q$ acts on the first factor by translations. Now using (5.7) we see that $(H^i_c(f^{-1}s))_\psi = 0$ if $i \neq 2d$. So we have proved (i) and also that the sheaf $(R^{2d}f_!\overline{\mathbb{Q}}_\ell)_\psi$ is supported on $\mathcal{B}_A \times \underline{\underline{t}}'_0$. We also note that $H^{2d}_c(f^{-1}s)$ (where $s \in \mathcal{B}_A \times \underline{\underline{t}}'_0$) is isomorphic to a direct sum of q copies of $H^{2d}_c(A^d)$ and $\mathbb{F}_q$ permutes these factors. So $(H^{2d}_c(f^{-1}s))_\psi$ is isomorphic to $\overline{\mathbb{Q}}_\ell(-d)$.

In order to finish the proof we have to show that in fact we have a constant sheaf on $\mathcal{B}_A \times \underline{\underline{t}}'_0$. That it is the constant sheaf $\overline{\mathbb{Q}}_\ell(-d)$ will then follow from the last sentence. This will result from two applications of Base Change (5.2).

Let $Z = f^{-1}(\mathcal{B}_A \times \underline{\underline{t}}'_0) \subset \mathcal{Y}$. We have a commutative diagram

$$\begin{array}{ccc} Z & \xrightarrow{\ i\ } & \mathcal{Y} \\ {\scriptstyle f|Z}\downarrow & & \downarrow{\scriptstyle f} \\ \mathcal{B}_A \times \underline{\underline{t}}'_0 & \xrightarrow[\ i\ ]{} & \mathcal{B} \times \underline{\underline{t}}'_0 \end{array}$$

We also have $Z = \{(x,gT,A') \in \mathcal{Y} \mid (\mathrm{Ad}g^{-1})A \in \underline{b},\ x^q - x = 0\}$. Thus $Z \cong \mathbb{F}_q \times \rho^{-1}(\mathcal{B}_A) \times \underline{\underline{t}}'_0$, where $\rho\colon G/T \to \mathcal{B}$ is the map $gT \to gBg^{-1}$. By (5.2) it is sufficient to show that $(R^{2d}(f|Z)_!\overline{\mathbb{Q}}_\ell)_\psi$ is a constant sheaf on $\mathcal{B}_A \times \underline{\underline{t}}'_0$. Since we also have the commutative diagram

$$\begin{array}{ccccc} Z & \xrightarrow{\text{projection}} & \rho^{-1}(\mathcal{B}_A) \times \underline{\underline{t}}'_0 & \longrightarrow & G/T \\ & & {\scriptstyle \rho}\downarrow & & \downarrow{\scriptstyle \rho} \\ & & \mathcal{B}_A \times \underline{\underline{t}}'_0 & \longrightarrow & \mathcal{B} \end{array}$$

we are reduced to showing that if we consider the map $\rho\colon G/T \to \mathcal{B}$, the sheaf $R^{2d}\rho_!\bar{\mathbb{Q}}_\ell$ is a constant sheaf on $\mathcal{B}$. Note that the natural map $\phi: G/T \to G/B$ is locally trivial, i.e., there is a covering of G/B by (Zariski) open sets O_i such that $\phi^{-1}(O_i) \cong O_i \times \mathbb{A}^d$; this follows from the cellular decomposition of G/B (see [5], p. 347). Hence ρ is locally trivial. From this it follows (again by Base Change) that there is a covering of $\mathcal{B}$ by Zariski open sets such that $R^{2d}\rho_!\bar{\mathbb{Q}}_\ell$ is a constant sheaf on each open set, i.e., that $R^{2d}\rho_!\bar{\mathbb{Q}}_\ell$ is a locally constant sheaf on $\mathcal{B}$. Now it is a deep result that if X is a connected scheme and $\bar{x}$ is a geometric point of X, the category of locally constant constructible sheaves of $\mathbb{Z}/\ell^n\mathbb{Z}$-modules on X is equivalent to the category of finitely-generated $\mathbb{Z}/\ell^n\mathbb{Z}$-modules on which the fundamental group $\pi_1(X,\bar{x})$ acts continuously (see [63], p. 82, 2.3; [56], [60]). The fundamental group of $\mathcal{B}$ is trivial since it is a rational projective variety (see [60], p. 285). Thus $R^{2d}\rho_!\bar{\mathbb{Q}}_\ell$ is a constant sheaf on $\mathcal{B}$. This proves the theorem.

<u>Remark</u>. Michael Artin has pointed out to me that there is a simpler way of proving that $R^{2d}\rho_!\bar{\mathbb{Q}}_\ell$ is a constant sheaf on $\mathcal{B}$, once we have shown that it is locally constant. This follows from the fact if X is a connected <u>normal</u> variety and Φ is a sheaf on X which is constant on some non-empty open set U of X then Φ is a constant sheaf on X.

We now consider the composite morphism $\pi = \sigma f: \mathcal{Y} \to \underline{\underline{t}}_0'$, and use the Grothendieck spectral sequence (see 5.6). We regard $\mathbb{F}_q$ as acting trivially on $\underline{\underline{t}}_0'$; then π commutes with the actions of $\mathbb{F}_q$ on $\mathcal{Y}$ and $\underline{\underline{t}}_0'$. Thus we can consider the sheaf $(R^i\pi_!\bar{\mathbb{Q}}_\ell)_\psi$, the part of $R^i\pi_!\bar{\mathbb{Q}}_\ell$ on which $\mathbb{F}_q$ acts according to Ψ.

Theorem 7.3 ([75], 4.2). The sheaf $(R^i\pi_!\bar{\mathbb{Q}}_\ell)_\psi$ is the constant sheaf $H_c^{i-2d}(\mathcal{B}_A)(-d)$ on $\underline{\underline{t}}_0'$, for each i. The action of W on $\mathcal{Y}$ gives rise to a representation of W on $H_c^i(\mathcal{B}_A)$.

Proof. By the Grothendieck spectral sequence (5.6), using the fact (7.2(i)) that $(R^jf_!\bar{\mathbb{Q}}_\ell)_\psi = 0$ if $j \neq 2d$, we get $(R^i\sigma_!)(R^{2d}f_!\bar{\mathbb{Q}}_\ell)_\psi \cong (R^{i+2d}\pi_!\bar{\mathbb{Q}}_\ell)_\psi$. Let Φ denote the sheaf on supported by $\mathcal{B}_A$, which is the constant sheaf $\bar{\mathbb{Q}}_\ell(-d)$ on $\mathcal{B}_A$.

We have a commutative diagram

$$\begin{array}{ccc} \mathcal{B}\times \underline{\underline{t}}_0' & \xrightarrow{\ \pi_1\ } & \mathcal{B} \\ \downarrow{\scriptstyle\sigma} & & \downarrow \\ \underline{\underline{t}}_0' & \longrightarrow & \mathrm{Spec}\ K \end{array}$$

where π_1 is projection. Then Theorem 7.2 (ii) implies that $\pi_1^*\Phi = (R^{2d}f_!\bar{\mathbb{Q}}_\ell)_\psi$. By Base Change (5.2) it follows that $R^i\sigma_!(\pi_1^*\Phi)$ is a constant sheaf on $\underline{\underline{t}}_0'$, and thus $(R^{i+2d}\pi_!\bar{\mathbb{Q}}_\ell)_\psi$ is a constant sheaf on $\underline{\underline{t}}_0'$, as required. The stalk at a geometric point of $\underline{\underline{t}}_0'$ centered at $A' \in \underline{\underline{t}}_0'$ of the sheaf $R^i\sigma_!(\pi_1^*\Phi)$ is isomorphic to $H_c^i(\sigma^{-1}A', \pi_1^*\Phi)$, which in turn is isomorphic to $H_c^i(\mathcal{B}_A)(-d)$. This proves the first statement of the theorem. The action of W on $\mathcal{Y}$ gives rise to an action of W on the constant sheaf $(R^{i+2d}\sigma_!\bar{\mathbb{Q}}_\ell)_\psi$. Thus we get an action of W on $H_c^i(\mathcal{B}_A)(-d)$, i.e., a representation of W on $H_c^i(\mathcal{B}_A)$, since $H_c^i(\mathcal{B}_A)$ is isomorphic to $H_c^i(\mathcal{B}_A)(-d)$. This proves the theorem.

Next we make the connection between these representations and the trigonometric sums $S(A,A')$. First we recall that in defining $f\colon \mathcal{Y} \to \mathcal{B}\times t_0'$ by $f(x,gT,A') = (gBg^{-1},A')$, we have made a choice of $B \in \mathcal{B}$. So we now write f_B for f. Then $f_{\dot{w}B\dot{w}^{-1}} \cdot w = (i,w) \cdot f_B$ $(w \in W)$. Denote by α_B^i the isomorphism of constant sheaves $(R^i\pi_!\overline{\mathbb{Q}}_\ell)_\psi \tilde{\to} H_c^{i-2d}(\mathcal{B}_A)(-d)$ obtained in the proof of (7.3). Let $w \to \rho^i(w)$ be the action of W on the constant sheaf $(R^i\pi_!\overline{\mathbb{Q}}_\ell)_\psi$. Then $\alpha^i_{\dot{w}B\dot{w}^{-1}}\rho^i(w) = \alpha^i_B$ $(w \in W)$. Denote the representation of W on $H_c^i(\mathcal{B}_A)(-d)$, or on $H_c^i(\mathcal{B}_A)$, by r^i. Then we have $r^{i-2d} = \alpha^i_B\rho^i\alpha^{i^{-1}}_B$.

We now consider the action of F. We may suppose that $B = a^{-1}B_0a$, $T = a^{-1}T_0a$, for some $a \in G$. Then we have an isomorphism $n \to a^{-1}na$ of $N(T_0)$ onto $N(T)$, which induces an isomorphism γ of $W(T_0)$ onto $W(T)$. Now T corresponds to $w_1 \in W(T_0)$ (in the sense of (2.8)) where $\dot{w}_1 = a(Fa)^{-1} \in N(T_0)$. We have $FB = (Fa)^{-1}aBa^{-1}(Fa) = \dot{w}_2B\dot{w}_2^{-1}$, where $\dot{w}_2 = a^{-1}\dot{w}_1a$ and thus w_2 corresponds to w_1 under γ.

<u>Remark</u>. It can be shown that the representations ρ^i of W are independent of the choice of B (see [75], 4.6).

We now denote the representation of $W(T_0)$ on $H_c^i(\mathcal{B}_A)$ obtained via the isomorphism γ, also by r^i. Then we can prove the following theorem.

<u>Theorem 7.4</u> ([75], 4.4). Let $w \in W(T_0)$ correspond to the torus T. Then $S(A,A') = q^d \sum_{i\geq 0} (-1)^i \mathrm{Tr}(F\cdot r^i(w)^{-1}, H_c^i(\mathcal{B}_A))$.

<u>Proof</u>. From the commutative diagram

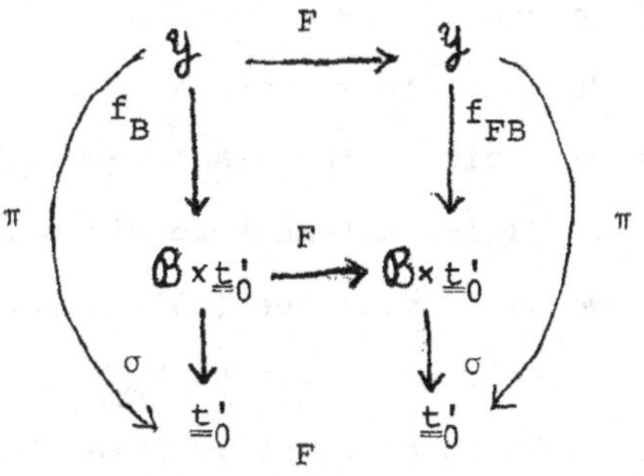

we get an isomorphism of constant sheaves $F^*: (R^i\pi_!\overline{Q}_\ell)_\psi \tilde{\to} (R^i\pi_!F^*\overline{Q}_\ell)_\psi$ on $\underline{\underline{t}}'_0$, where F^* denotes the pullback under F of the appropriate sheaves. Now $(R^i\pi_!\overline{Q}_\ell)_\psi$ is the constant sheaf $H_c^{i-2d}(\mathcal{B}_A)(-d)$ on $\underline{\underline{t}}'_0$, and it follows from the proofs of (7.2) and (7.3) that the morphism F^* is given by the endomorphism F of $H_c^{i-2d}(\mathcal{B}_A)(-d) \cong H_c^{i-2d}(\mathcal{B}_A,\overline{Q}_\ell(-d))$ which arises from the endomorphism F of $\mathcal{B}_A$. If we now make an identification of $H_c^{i-2d}(\mathcal{B}_A)(-d)$ with $H_c^{i-2d}(\mathcal{B}_A)$, then we have to replace the endomorphism F of the former group by the endomorphism q^dF of $H_c^{i-2d}(\mathcal{B}_A)$. Hence we have a commutative diagram

$$\begin{array}{ccccc}
(R^i\pi_!\overline{Q}_\ell)_\psi & \xrightarrow{\alpha^i_{FB}} & H_c^{i-2d}(\mathcal{B}_A)(-d) & \tilde{\to} & H_c^{i-2d}(\mathcal{B}_A) \\
\big\downarrow F^* & & & & \big\downarrow q^dF \\
(R^i\pi_!F^*\overline{Q}_\ell)_\psi & \xrightarrow[\alpha^i_B]{} & H_c^{i-2d}(\mathcal{B}_A)(-d) & \tilde{\to} & H_c^{i-2d}(\mathcal{B}_A)
\end{array}$$

from which we get $\alpha_B^i F^* = q^d F \alpha_{FB}^i = q^d F \alpha_B^i \rho^i(w)^{-1} = q^d F r^{i-2d}(w)^{-1}\alpha_B^i$, where $w \in W$ is given by $FB = \dot{w}B\dot{w}^{-1}$.

Now consider the fibre over $A' \in \underline{\underline{t}}_0'$ of π. We remark that G/T is isomorphic to the $(Ad'G)$-orbit of A' since A' is strongly regular. (This statement requires some proof, using [5], 6.7, which we will omit; see [75], 2.10.) Thus the fibre of π over A' is precisely the variety $Y_{A,A'}$ defined earlier. Hence we get that the stalk at a geometric point of $\underline{\underline{t}}_0'$ centered at A' of the sheaf $(R^i\pi_!\bar{\mathbb{Q}}_\ell)_\psi$ is isomorphic to $H_c^i(Y_{A,A'})_\psi$, and thus $H_c^i(Y_{A,A'})_\psi \cong H_c^{i-2d}(\mathcal{B}_A)(-d)$. Theorem 7.4 then follows from (7.1).

<u>Remarks</u>.

1. The variety $\mathcal{B}_A$ is in general singular. For example, if $G = GL_3$, $A = \begin{pmatrix} 0 & 1 & 0 \\ 0 & 0 & 0 \\ 0 & 0 & 0 \end{pmatrix}$, $\mathcal{B}_A$ is the union of two intersecting projective lines over K:

If $G = Sp_4$ and A is the nilpotent element which has two blocks $\begin{pmatrix} 0 & 1 \\ 0 & 0 \end{pmatrix}$ along the diagonal and zeros everywhere else, $\mathcal{B}_A$ is the union of three projective lines over K:

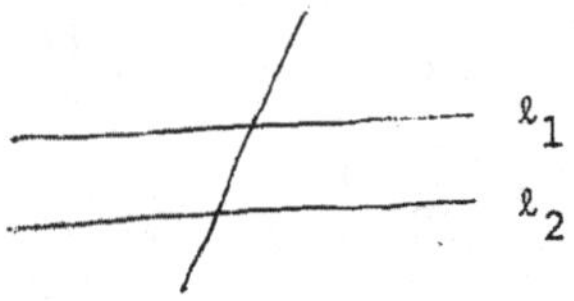

In this case we have a standard Frobenius action on $\mathcal{B}_A$ stabilizing each line, and a twisted Frobenius action on $\mathcal{B}_A$ interchanging ℓ_1 and ℓ_2. This corresponds to the fact that on the two unipotent classes in the finite group $Sp(4,q)$ with representatives $\begin{pmatrix} 1 & 1 & 0 & 0 \\ 0 & 1 & 0 & 0 \\ 0 & 0 & 1 & -1 \\ 0 & 0 & 0 & 1 \end{pmatrix}$ and $\begin{pmatrix} 1 & 1 & 0 & 0 \\ 0 & 1 & 0 & 0 \\ 0 & 0 & 1 & -\gamma \\ 0 & 0 & 0 & 1 \end{pmatrix}$ (where γ is a generator of $\mathbb{F}_q^*$) the principal series irreducible characters have vales $1 + 3q$ and $1 + q$, which are equal to $|\mathcal{B}_A^F|$ in the two cases (see [76], p. 517).

Since $\mathcal{B}_A$ is singular we cannot use the deep theorems of Deligne (see [68]) on the eigenvalues of F on the groups $H_c^i(\mathcal{B}_A)$. However, recently P. Slodowy in Bonn has proved that there is a non-singular variety whose cohomology is the same as that of $\mathcal{B}_A$, and from this it follows that the eigenvalues of F on $H_c^i(\mathcal{B}_A)$ have absolute value $q^{i/2}$.

For various results on the variety $\mathcal{B}_A$ the reader is referred to [80], [83].

2. Let $Z = C_G(A)$. Then Z acts on $\mathcal{Y}$ by $z(x,gT,A') = (x,zgT,A')$ $(z \in Z)$ and this action commutes with the action of W. Thus Z acts on the constant sheaf $(R^i\pi'\overline{\mathbb{Q}}_\ell)_\psi$ on $\underline{\underline{t}}_0'$ and we get a representation of Z on $H_c^i(\mathcal{B}_A)$ which commutes with the representation r^i of W. Now Z^0 acts trivially on $H_c^i(\mathcal{B}_A)$, using (6.5), and thus the finite group $C = Z/Z^0$ acts on $H_c^i(\mathcal{B}_A)$. Springer ([75], 6.10) shows (assuming q sufficiently large) that if ζ is an irreducible representation of C then the representation of W on the ζ-isotypic component of the top cohomology group $H_c^{2e}(\mathcal{B}_A)$, where $e = \dim \mathcal{B}_A$, is irreducible and each irreducible representation of W is obtained exactly once as

we vary A over a set of representatives for the (AdG)-orbits of nilpotent elements in $\underline{g}$ and ζ over a set of representatives for the isomorphism classes of irreducible representations of C. Shoji [70] has studied this connection between nilpotent elements in $\underline{g}$ and representations of W when G is classical or of type F_4, and Hotta and Springer [37] when $G = GL_n$ or when A is of "parabolic type."

Application to Green functions.

Given the F-stable maximal torus T, we fix a strongly regular element A' in $\underline{t}'^F$, and define the following function on the nilpotent elements of $\underline{g}^F$.

Definition 7.5. $Q_{T,\underline{g}}(A) = \varepsilon_G \varepsilon_T q^{-d} S(A,A')$.

We state without proof certain orthogonality relations satisfied by the functions $Q_{T,\underline{g}}$ which are similar to the relations of (6.15). We note that we are justified in our notation $Q_{T,\underline{g}}$ since (7.4) shows that S(A,A') is independent of the choice of A'.

Theorem ([75], 5.6). Let T,T' be two F-stable maximal tori in G. Then (for sufficiently large q) we have

$$(7.6) \qquad \frac{1}{|G^F|} \sum_{\substack{X \in \underline{g}^F \\ X \text{ nilpotent}}} Q_{T,\underline{g}}(X) Q_{T',\underline{g}}(X) = \frac{|N(T,T')^F|}{|T^F|\,|T'^F|}.$$

We will now sketch a proof that if $G = GL_n$ and F is the standard Frobenius, so that $G^F = GL(n,q)$, then $Q_{T,}(A)$ is a polynomial in q. We can assume A is in Jordan form. The variety $\mathcal{B}$ can be identified with the variety of complete flags

$$0 \subset V_1 \subset V_2 \subset \cdots \quad V_n = V, \ \dim V_{i+1}/V_i = 1,$$

where V is the vector space over K on which G acts. Then $\mathcal{B}_A$ is the subvariety of flags such that V_{i+1}/V_i is annihilated by A. Let $\mathcal{P}_A$ be the variety of lines (i.e., one-dimensional subspaces) of V annihilated by A. Then $\mathcal{P}_A \cong \mathbb{P}(M)$, where $M = \ker A$.

We have a map $\pi\colon \mathcal{B}_A \to \mathcal{P}_A$ which takes the flag $0 \subset V_1 \cdots$ to V_1. If $v \in M$, A induces a nilpotent transformation $\bar{A}$ on $M/\langle v\rangle$, so $\bar{A}$ can be regarded as an element of the Lie algebra of GL_{n-1}. The fibre of π over $\langle v\rangle$ is isomorphic to $\mathcal{B}_{\bar{A}}$. We have a filtration $M = M_0 \supset M_1 \supset \cdots \supset M_d = 0$, where $M_i = M \cap \operatorname{Im} A^i$. For all $v \in M_i - M_{i+1}$, the fibres of π are isomorphic as varieties over $\mathbb{F}_q$. Let $Y_i = \pi^{-1}(\mathbb{P}(M_i) - \mathbb{P}(M_{i+1}))$ (making an identification of $\mathcal{P}_A$ with $\mathbb{P}(M)$). Then one can show (see [71] for details) that the map $\pi\colon Y_i \to \mathbb{P}(M_i) - \mathbb{P}(M_{i+1})$ is locally trivial, and thus locally $H^i_c(Y_i)$ looks like $\bigoplus_{j+k=i} \{H^j_c(\mathbb{P}(M_i) - \mathbb{P}(M_{i+1})) \otimes H^k_c(\mathcal{B}_{\bar{A}})\}$. Now $\mathbb{P}(M_i) - \mathbb{P}(M_{i+1})$ is a union of affine spaces, and by induction we can assume that the eigenvalues of F on $H^k_c(\mathcal{B}_{\bar{A}})$ are integral powers of q. Furthermore the representations of W ($\cong S_n$ in this case) are integral. Using Theorem 7.4 we see that $Q_{T'}(A)$ is a polynomial in q with integer coefficients.

The fibration of $\mathcal{B}_A$ described above has been given by Spaltenstein in [loc. cit.]. Similar results for classical groups, showing that the $Q_{T,\underline{g}}(A)$ are polynomials in q also in that case, can be found in [77].

We would now like to connect the functions $Q_{T,\underline{g}}$ on the nilpotent elements of $\underline{g}^F$ with the Green functions Q_T^G on the unipotent elements of G introduced in Chapter VI. We describe the work of Kazhdan [41] which gives this connection. We start with a brief description of the Kirillov theory of orbits for nilpotent Lie algebras over $\mathbb{R}$ (see e.g., [43], p. 87).

Let N be a connected, simply connected, nilpotent Lie group over $\mathbb{R}$ and $\underline{n}$ its Lie algebra. Then $\underline{n}$ is a nilpotent Lie algebra over $\mathbb{R}$. Let $\underline{n}'$ be the dual of $\underline{n}$, i.e., $\underline{n}' = \mathrm{Hom}_{\mathbb{R}}(\underline{n},\mathbb{R})$.

<u>Definition</u>. Let $\lambda \in \underline{n}'$. Then B_λ is the bilinear alternating form on $\underline{n}$ given by $B_\lambda(x_1,x_2) = \lambda[x_1,x_2] \quad (x_1,x_2 \in \underline{n})$.

<u>Definition</u>. A subalgebra $\underline{h}$ of $\underline{n}$ which is a maximal isotropic subspace for B_λ is called a polarization for λ.

For a general discussion of polarizations the reader is referred to [20]. Such subalgebras exist in the nilpotent case and the following construction is due to M. Vergne. Let $\underline{n} = \underline{n}_0 \supset \underline{n}_1 \supset \cdots \supset \underline{n}_k = 0$ be a series of subspaces of $\underline{n}$ such that $\dim \underline{n}_i/\underline{n}_{i+1} = 1$. Let $\underline{h}_i$ be the kernel of the restriction of B_λ to $\underline{n}_i$, and let $\underline{h} = \sum_{i=1}^{k} \underline{h}_i$. Then $\underline{h}$ is a maximal isotropic subspace of B_λ. If the $\underline{n}_i$ are also ideals, then $\underline{h}$ is a subalgebra. Hence we can always find polarizations for λ if

$\underline{\underline{n}}$ is nilpotent. The idea in finding polarizations is that λ is only a linear functional on $\underline{\underline{n}}$, but it is a representation of $\underline{\underline{h}}$. So we are finding a maximal subalgebra $\underline{\underline{h}}$ of $\underline{\underline{n}}$ which has λ as a representation. (Note, however, that $\underline{\underline{h}}$ is not uniquely determined by λ.)

Now let $H = \exp \underline{\underline{h}}$, and define a character of H by $\chi_\lambda(\exp x) = e^{i\lambda(x)}$ $(x \in \underline{\underline{h}})$. Let $\rho(\lambda,\underline{\underline{h}}) = \mathrm{Ind}_H^N(\chi_\lambda)$. Then Kirillov showed that the representations $\rho(\lambda,\underline{\underline{h}})$ of N are irreducible and that there is a bijection $\underline{\underline{n}}'/N \to \varepsilon(N)$, where $\varepsilon(N)$ is the set of unitary equivalence classes of irreducible unitary representations of N.

We now return to our group G and give a similar description of the characters of $U^F \subset G^F$, where U is an F-stable maximal unipotent subgroup of G, provided p is sufficiently large. Let V be the variety of unipotent elements of G and $\underline{\underline{V}}$ the variety of nilpotent elements of . We assume, for the rest of this chapter, that p is large enough so that the maps $\exp\colon \underline{\underline{V}} \to V$ and $\ell n\colon$ $V \to \underline{\underline{V}}$ are defined as in characteristic 0 and the Campbell-Hausdorff formulas (see e.g., [34]) hold, and that the Killing form is non-degenerate on the Lie algebras of reductive subgroups of G. (For example if $G = GL_n$, $p > n$ will do.)

Let $\underline{\underline{n}} \subset \underline{\underline{g}}$ be the Lie algebra of U, $\underline{\underline{n}}'$ the linear dual (over K) of $\underline{\underline{n}}$. U acts on $\underline{\underline{n}},\underline{\underline{n}}'$ by Ad,Ad' respectively, and F acts on $\underline{\underline{n}}$ and $\underline{\underline{n}}'$. Let $\lambda \in \underline{\underline{n}}'^F$. We find a polarization

$\underline{\underline{h}} \subset \underline{\underline{n}}$ for λ, just as in the real case. In other words we define B_λ as before and imitate the construction given to find a polarization. Then we have a corresponding subgroup $H = \exp \underline{\underline{h}}$ of U, which is fixed by F.

We fix a non-trivial character $\psi: \mathbb{F}_q \to \overline{\mathbb{Q}}_\ell^*$. Let the character $\phi_\lambda: H^F \to \overline{\mathbb{Q}}_\ell^*$ be given by $\phi_\lambda(\exp x) = \psi(\lambda(x))$ $(x \in \underline{\underline{h}})$. Then define $\chi_\lambda = \mathrm{Ind}_{H^F}^{U^F}(\phi_\lambda)$, so that χ_λ is a character of U^F.

Theorem 7.7 ([41], Propositions 1 and 2).

(i) $\chi_\lambda(v) = q^{-\dim \Omega_\lambda/2} \sum_{\mu \in \Omega_\lambda^F} \psi(\mu(\ell n\ v))$ $(v \in U^F)$,

where $\Omega_\lambda = (\mathrm{Ad}\ U)\lambda$ is the U-orbit of λ.

(ii) χ_λ is irreducible for some λ.

(iii) $\chi_\lambda = \chi_{\lambda'}$ if and only if $\Omega_\lambda = \Omega_{\lambda'}$.

(iv) Every irreducible character of U^F is equal to χ_λ

(v) $\dim \chi_\lambda = |\Omega_\lambda^F|^{1/2} = q^{\dim \Omega_\lambda/2}$.

Proof (i). Let $v \in U^F$, $v = \exp w$ where $w \in \underline{\underline{n}}$. Then, by definition,

$$\chi_\lambda(v) = \frac{1}{|H^F|} \sum_{u \in U^F} a_\lambda((\mathrm{Ad} u)w), \text{ where}$$

$$a_\lambda(x) = \begin{cases} \psi(\lambda(x)), & \text{if } x \in \underline{\underline{h}}^F \\ 0, & \text{otherwise.} \end{cases}$$

Let $\underline{\underline{\ell}}$ be the affine subspace of $\underline{\underline{n}}'$ consisting of all linear functionals ν such that $\nu|\underline{\underline{h}} = \lambda$, i.e., $\underline{\underline{\ell}} = \underline{\underline{h}}^\perp + \lambda$ where $\underline{\underline{h}}^\perp$ is the annihilator of $\underline{\underline{h}}$ in $\underline{\underline{n}}'$. Then we have $a_\lambda(x) = \frac{1}{|\underline{\underline{\ell}}^F|} \sum_{\nu \in \underline{\underline{\ell}}^F} \psi(\nu(x))$: for if $x \in \underline{\underline{h}}^F$, then $\nu(x) = \lambda(x)$ for all ν

in $\underline{\underline{\ell}}^F$ and we just get $\psi(\lambda(x))$, and if $x \notin \underline{\underline{h}}^F$ then as ν varies over $\underline{\underline{h}}^{\perp} + \lambda$ we get $\sum_{\nu} \psi(\nu(x)) = 0$ since ψ is a non-trivial character of $\mathbb{F}_q$. Hence we get

$$\chi_\lambda(v) = \frac{1}{|H^F|}\frac{1}{|\underline{\underline{\ell}}^F|} \sum_{\substack{u\in U^F \\ \nu\in \underline{\underline{\ell}}^F}} \psi(\nu(\mathrm{Ad}\ u)(w))$$

$$= \frac{1}{|H^F|}\frac{1}{|\underline{\underline{\ell}}^F|} \sum_{u,\nu} \psi(((\mathrm{Ad}'u^{-1})\nu)(w)).$$

Next, we see that $\underline{\underline{\ell}}$ is (Ad H)-stable, and we show that in fact H acts transitively on $\underline{\underline{\ell}}$. Let $H_0 \subset H$ be the stabilizer of λ in U, and let $\underline{\underline{h}}_0$ be the kernel of the form B_λ on $\underline{\underline{n}}$, i.e., $\underline{\underline{h}}_0 = \{x \in \underline{\underline{n}} \mid \lambda[x,y] = 0 \text{ for all } y \text{ in } \underline{\underline{n}}\}$. Now (transferring exp to ad $\underline{\underline{n}}$) we have $\lambda(\mathrm{ad}\ x)y) = 0$ for all y in $\underline{\underline{n}}$ if and only if $\lambda(\exp \mathrm{ad}\ x)(y) = \lambda(y)$ for all y in $\underline{\underline{n}}$. Just as in characteristic 0 (see, e.g., [34], p. 116, Ex. 3) it follows that $H_0 = \exp \underline{\underline{h}}_0$ and, in particular, H_0 is connected.

By Witt's Theorem we can write $\underline{\underline{n}} = \underline{\underline{h}}_0 \oplus \underline{\underline{a}} \oplus \underline{\underline{a}}'$ where $\underline{\underline{h}} = \underline{\underline{h}}_0 \oplus \underline{\underline{a}}$ and $\dim \underline{\underline{a}} = \dim \underline{\underline{a}}'$. So $\mathrm{codim}\ \underline{\underline{h}} = \dim H - \dim H_0$, $\dim \Omega_\lambda = \dim U - \dim H_0 = 2\ \mathrm{codim}\ \underline{\underline{h}}$, $\dim \underline{\underline{\ell}} = \dim \underline{\underline{h}}^{\perp} = \mathrm{codim}\ \underline{\underline{h}}$ $= \frac{1}{2} \dim \Omega_\lambda$. Consider the map $\sigma\colon H_0\backslash H \to \underline{\underline{\ell}}$ given by $\sigma(x) = (\mathrm{Ad}\ x)\lambda$. This is injective and the above remarks on dimensions show that σ is surjective. Thus H acts transitively on $\underline{\underline{\ell}}$. Now the H^F-orbits in $\underline{\underline{\ell}}$ are classified by the F-conjugacy classes of $H_0/(H_0)^0$, as in (2.7). Since H_0 is connected, we also have that H^F is transitive on $\underline{\underline{\ell}}$.

Now it follows that $\chi_\lambda(v) = \frac{1}{|\underline{\underline{\ell}}^F|} \sum_{\nu\in\Omega_\lambda^F} \psi(\nu(w))$, i.e.,

$\chi_\lambda(v) = q^{-\dim \Omega_\lambda/2} \sum_{\nu\in\Omega_\lambda^F} \psi(\nu(w))$, where $w = \ell n\ v$, which proves (i).

(ii) We compute $(\chi_\lambda, \chi_\lambda)$ using (i). We get

$$(\chi_\lambda, \chi_\lambda) = q^{-\dim \Omega_\lambda} \frac{1}{|\underline{\underline{n}}^F|} \sum_{x\in\underline{\underline{n}}^F} |\sum_\mu \psi(\mu(x))|^2,$$

where μ runs over $|\Omega_\lambda^F|$ linear maps of the F_q-vector space $\underline{\underline{n}}^F$. Now $|\Omega_\lambda^F| = \frac{|U^F|}{|H_0^F|} = q^{\dim \Omega_\lambda}$ since U is unipotent and hence $H_0\backslash U$ is isomorphic to affine space (see e.g. [19], p. 535). This proves that $(\chi_\lambda, \chi_\lambda) = 1$.

(iii) is proved analogously to (ii) since we can compute $(\chi_\lambda, \chi_{\lambda'})$ in the same way. Also (v) follows from (i).

(iv) We have $\sum_\lambda (\dim \chi_\lambda)^2 = \sum_\lambda |\Omega_\lambda^F| = q^{\dim \underline{\underline{n}}} = |U^F|$, where λ runs over a set of representatives from the Ad U-orbits on $\underline{\underline{n}}'$. This proves (iv), and hence the theorem.

As before let T be an F-stable maximal torus of G and let $\underline{\underline{t}} \subset \underline{\underline{g}}$ be the Lie algebra of T. We identify $\underline{\underline{t}}$ and its dual $\underline{\underline{t}}'$ by means of the Killing form. Let A' be a strongly regular semisimple element in $\underline{\underline{t}}^F$: i.e. $C_G(A') = T$. Let $\Omega(A') = (\mathrm{Ad}\ G)A'$ be the G-orbit of A'. Since $C_G(A')$ is connected, we see as in the proof of (2.7) that $\Omega(A')^F = (\mathrm{Ad}\ G^F)A'$. Now we can write the function $Q_{T,\underline{\underline{g}}}$ on $\underline{\underline{V}}^F$ defined in (7.5) as

$$Q_{T,\underline{\underline{g}}}(x) = \frac{\varepsilon_G \varepsilon_T}{|U^F|} \sum_{y\in\Omega(A')^F} \psi(\langle x,y\rangle) \qquad (x \in \underline{\underline{V}}),$$

where $< \ , \ >$ denotes the Killing form. We define a function Φ on V^F as follows:

$$\Phi(v) = \varepsilon_G \varepsilon_T \, Q_{T,\underline{g}}(\ell n \ v).$$

The following theorem is a crucial result in Kazhdan's paper.

Theorem 7.8 ([41], Theorem 1). The function Φ is a (proper) character of U^F.

Proof. By (7.7) it is sufficient to show that the scalar product of Φ with any χ_λ is a non-negative integer. Since $\chi_\lambda = \mathrm{Ind}_{H^F}^{U^F}(\phi_\lambda)$ for some $\lambda \in \underline{n}'^F$, we have

$$\begin{aligned}
(\Phi,\chi_\lambda)_{U^F} &= (\Phi,\phi_\lambda)_{H^F} \\
&= \frac{1}{|H^F|} \sum_{u \in H^F} \varepsilon_G \varepsilon_T \, Q_{T,\underline{g}}(\ell n \ u) \, \overline{\phi_\lambda(u)} \\
&= \frac{1}{|H^F|} \sum_{x \in \underline{h}^F} \varepsilon_G \varepsilon_T Q_{T,\underline{g}}(x) \ \psi(-\lambda(x)) \\
&= \frac{1}{|H^F|} \frac{1}{|U^F|} \sum_{\substack{y \in \Omega(A')^F \\ x \in \underline{h}^F}} \psi\{\langle x,y\rangle - \lambda(x)\}. \\
&= \frac{1}{|U^F|} |X^F|,
\end{aligned}$$

where $X \subset \Omega(A')$ is the variety of all $y \in \Omega(A')$ such that the linear functional $x \to \langle x,y\rangle$ on $\underline{h}$ coincides with λ. So in order to prove the theorem we have to show that $|U^F|$ divides $|X^F|$.

The variety X has been defined with reference to the strongly regular element A' whose centralizer is T and the F-stable unipotent subgroup U. Keeping U fixed, we conjugate T

over K and assume now that T normalizes U. We can no longer assume T (or A') F-stable. Thus we will now essentially consider a variety which is isomorphic over K to X, and define a partition of this variety into locally closed subvarieties.

By the Bruhat decomposition we have $G = \bigcup_w T U \dot{w} U_w^- = \bigcup_w U \dot{w} U_w^- T$, where w runs over W(T). Let $g \in G_w = U\dot{w}U_w^- T$, and consider $(\mathrm{Ad}\ g)A'$. We have $(\mathrm{Ad}\ T)A' = A'$, $(\mathrm{Ad}\ U_w^-)A' = A' + \underline{\underline{n}}_w^-$, where $\underline{\underline{n}}_w^-$ is the Lie algebra of U_w^-. Then $(\mathrm{Ad}\ U)(\mathrm{Ad}\ \dot{w})(A' + \underline{\underline{n}}_w^-)$ $= (\mathrm{Ad}\ U)((\mathrm{Ad}\ w)A' + (\mathrm{Ad}\ \dot{w})\underline{\underline{n}}_w^-)$. Since $(\mathrm{Ad}\ U)(\mathrm{Ad}\ \dot{w})A' \in \underline{\underline{t}} \oplus \underline{\underline{n}}$, and $\underline{\underline{t}} \oplus \underline{\underline{n}}$ is orthogonal to $\underline{\underline{n}}$ in the Killing form we see that we have to consider functionals induced on $\underline{\underline{n}}$ (via the Killing form) by elements of $(\mathrm{Ad}\ U)(\mathrm{Ad}\ \dot{w})\underline{\underline{n}}_w^- \subset$. Now $\underline{\underline{n}} = \underline{\underline{n}}_w^- \oplus \underline{\underline{n}}_w$, where $\underline{\underline{n}}_w$ is the Lie algebra of $U_w = U \cap \dot{w}U\dot{w}^{-1}$. So the linear functionals induced on $\underline{\underline{n}}$ by $(\mathrm{Ad}\ \dot{w})\underline{\underline{n}}_w^-$ are precisely those linear functionals on $\underline{\underline{n}}$ which vanish on $(\mathrm{Ad}\ \dot{w})\underline{\underline{n}}_w \subset \underline{\underline{n}}$, since the Killing form is (Ad G)-invariant. So we have to consider the variety $\bigcup_{w \in W} X_w$, where $X_w = \{(\mu,u) \in \underline{\underline{n}}' \times U \mid \mu|(\mathrm{Ad}\ \dot{w})\underline{\underline{n}}_w = 0,\ (\mathrm{Ad}\ u)\mu|\underline{\underline{h}} = \lambda\}$.

We can now describe our situation as follows. Suppose we have two subalgebras $\underline{\underline{\ell}}, \underline{\underline{h}}$ of $\underline{\underline{n}}$ and corresponding subgroups $L = \exp \underline{\underline{\ell}}$, $H = \exp \underline{\underline{h}}$ of U. Consider then the variety $Z \subset \underline{\underline{n}}' \times U$ defined by $Z = \{(\mu,u) \mid \mu|\underline{\underline{\ell}} = 0,\ (\mathrm{Ad}\ u)\mu|\underline{\underline{h}} = \lambda\} =$ $\{(\mu,u) \mid \mu \in \underline{\underline{\ell}}^{\perp} \cap (\mathrm{Ad}\ u)^{-1}(\underline{\underline{h}}^{\perp} + \lambda)\}$.

Consider the action of $H \times L$ on U given by $(h,\ell)u$ $= h^{-1}u\ell$. By a result of Rosenlicht [59] a quotient variety exists on an open set of U. Applying induction on the dimension to the complement we see that there is a partition of U into a finite number of locally closed subsets U_i such that the quotient space Y_i of U_i with respect to the action of $H \times L$ exists. Let $Z_i = Z \cap (\underline{\underline{n}}' \times U_i)$ and let $f\colon\ Z_i \to Y_i$ be given by $f(\mu,u) = \phi(u)$, where ϕ is the quotient morphism $U_i \to Y_i$.

Consider the fibres of f. We show that these fibres are either empty or isomorphic to affine space of dimension d $(= \dim U)$. Let $y \in Y_i$, $u \in \phi^{-1}(y)$. Then $\phi^{-1}(y)$, which is the double coset HuL, is isomorphic to the quotient space of $H \times (uL)$ by $uLu^{-1} \cap H$. Since a quotient space of a unipotent group is isomorphic to affine space ([19], p. 535) we see that $\phi^{-1}(y)$ is isomorphic to affine space of dimension $\ell + h - m$ where $\ell = \dim L$, $h = \dim H$, $m = \dim(uLu^{-1} \cap H)$. Let $\Psi = \underline{\underline{\ell}}^{\perp} \cap (\mathrm{Ad}\ u)^{-1}(\underline{\underline{h}}^{\perp} + \lambda)$. If Ψ is non-empty it is isomorphic to affine space of dimension $(d - \ell) + (d - h) - (d - m) = d + m - \ell - h$. We set up a map α: $\Psi \times \phi^{-1}(y) \to f^{-1}(Y_i)$ as follows: $\alpha(\mu,hu\ell) = ((\mathrm{Ad}\ \ell^{-1})\mu,\ hu\ell)\ (h\in H, \ell\in L)$. Then $\alpha(\mu,hu\ell) \in Z_i$, since $\mu \in \underline{\underline{\ell}}^{\perp}$ implies that $(\mathrm{Ad}\ \ell^{-1})\mu \in \underline{\underline{\ell}}^{\perp}$, and $(\mathrm{Ad}\ hu\ell)(\mathrm{Ad}\ \ell^{-1})\mu = (\mathrm{Ad}\ h)(\mathrm{Ad}\ u)\mu \in \underline{\underline{h}}^{\perp} + \lambda$ since $\underline{\underline{h}}^{\perp} + \lambda$ is Ad H-invariant. We can check that α is an isomorphism of $\Psi \times \phi^{-1}(y)$ onto $f^{-1}(y)$. Thus $f^{-1}(y)$, if non-empty, is isomorphic to affine space of dimension d.

Our original variety X is then isomorphic over K to a variety which is the disjoint union of locally closed subvarieties which are like the variety Z considered above. Thus Theorem 7.8 will follow from the following proposition.

Proposition 7.9 ([41], Proposition 3.) Let X_0 be a scheme over $\mathbb{F}_q$, and suppose $X = X_0 \underset{\mathbb{F}_q}{\times} \operatorname{Spec} K$. Suppose X is the disjoint union of locally closed subschemes X_i such that for each i there is a morphism $f_i \colon X_i \to Y_i$ whose fibres are either empty or isomorphic to $\mathbb{A}_d$. Then $|X^F|$ is divisible by q^d.

Proof. It is sufficient to show, by the Trace Formula (5.11) and the additivity property (6.1) that the eigenvalues of F on each $H_c^j(X_i)$ are divisible by q^d. (We note that the schemes X_i, and the morphisms f_i may not be defined over $\mathbb{F}_q$. But they are defined over some finite extension $\mathbb{F}_{q^n}$ of $\mathbb{F}_q$, and since the corresponding result about F^{q^n} would imply the result we want, we may as well assume that they are defined over $\mathbb{F}_q$.)

By the Leray spectral sequence (5.5), since the fibres of f_i are isomorphic to $\mathbb{A}_d$, we get $H_c^j(Y_i, R^{2d}f_i!\overline{\mathbb{Q}}_\ell) \cong H_c^{j+2d}(X_i)$, and by (5.3) the stalks of $R^{2d}f_i!\overline{\mathbb{Q}}_\ell$ at any geometric point of Y_i centered at a closed point are isomorphic to $\overline{\mathbb{Q}}_\ell(-d)$. Since F acts on $\overline{\mathbb{Q}}_\ell(-d)$ as multiplication by q^d, the proposition, and hence Theorem 7.8, follows from (5.12).

Remark. Deligne gives a proof of this proposition in a special case in [63], p. 175, as an application of cohomological methods.

We now denote the function $\Phi = \varepsilon_G \varepsilon_T Q_{T,\underline{\underline{g}}} \cdot \ell n$ on V^F by Φ_T^G. Thus we have a class function on the set of unipotent elements of G^F which is a character of U^F and we now extend it to a class function on all of G^F, bringing into play the characters of T^F. Let $g \in G^F$ and let $g = su$ be its Jordan decomposition. Let $\theta \in \hat{T}^F$. We define the function $X_{G,T,\theta}$ on G^F as follows.

Definition. $X_{G,T,\theta}(g) = 0$ if s is not conjugate to any element of T^F. Otherwise,

$$X_{G,T,\theta}(g) = \varepsilon_G \varepsilon_{C^0(s)} \frac{1}{|C^0(s)^F|} \sum_{\substack{x \in G^F \\ xsx^{-1} \in T^F}} \Phi^{C^0(s)}_{x^{-1}Tx}(u)\,\theta(xsx^{-1}).$$

Remarks. 1. $X_{G,T,\theta}(g)$ is an algebraic integer, for the term $\frac{1}{|C^0(s)^F|}$ will not appear if we sum only over representatives of conjugacy classes of tori in $C^0(s)$.

2. It turns out in the end that Φ^G_T is (up to sign) equal to Q^G_T and $X_{G,T,\theta}$ is the trace of $\varepsilon_G \varepsilon_T R^G_T(\theta)$.

It is our aim to show that $X_{G,T,\theta}$ is a virtual (generalized) character of G^F.

Definition. $\mathcal{H}$ is the set of connected reductive F-stable subgroups H of G such that H is the connected centralizer of some subset of T.

Definition. Define the Möbius function μ on $\mathcal{H}$ recursively by (i) $\mu(G) = 1$; (ii) If $C \in H$, $C \neq G$, $\sum_{\substack{C' \in \mathcal{H} \\ C' \supseteq C}} \mu(C') = 0$.

Definition. $K^G_T(\theta) = \sum_{H \in \mathcal{H}} \varepsilon_H \cdot \mu(H) \operatorname{Ind}^{G^F}_{H^F}(X_{H,T,\theta})$.

Using the formula for $X_{H,T,\theta}$ and properties of the Möbius function we can prove the following proposition. The proof is omitted.

Proposition 7.10 ([41], Proposition 4). The function $K_T^G(\theta)$ has its support on $Z(G^F) \times V^F$. In fact, we have

$$(7.11) \qquad (K_T^G(\theta))(su) = \begin{cases} \theta(s)(K_T^G(\theta))(u), & \text{if } s \in Z(G^F) \\ 0 & \text{, otherwise.} \end{cases}$$

Remark. Class functions on finite groups which are alternating sums of induced characters appear, for example, in the work of Dade (see [26], 33.8).

Next we state, also without proof, the following proposition. The main step in this proposition ([41], Proposition 6) relies essentially on a counting argument in the Tits building of G^F.

Proposition 7.12 ([41], Proposition 5). For every unipotent element u, $|Z(G^F)|(K_T^G(\theta)(u))/|C(u)^F|_p$, lies in the ring $\mathbb{Z}[q^{-1}]$.

Example. Consider the characters of $GL(2,q)$ given in Chapter III. If $\theta = \theta_{m,n}$, a character of the split torus T_0^F, we have $K_{T_0}^G(\theta) = \mathrm{Ind}_{T_0^F}^{G^F}(\theta) - \phi_{m,n}$ and takes the values $q^2 - 1$ and -1 at 1 and $\begin{pmatrix} 1 & 1 \\ 0 & 1 \end{pmatrix}$ respectively. A less trivial example is given by $G^F = GL(3,q)$, $T_0^F = \left\{ \begin{pmatrix} \gamma^a & 0 & 0 \\ 0 & \gamma^b & 0 \\ 0 & 0 & \gamma^c \end{pmatrix} \right\}$ where γ as before is a generator of $\mathbb{F}_q^*$. We define the following subgroups in $\mathcal{H}$:

$H_1 = T_0$, $H_2 = C_G\left(\begin{pmatrix} \gamma & 0 & 0 \\ 0 & \gamma & 0 \\ 0 & 0 & 1 \end{pmatrix}\right)$, $H_3 = C_G\left(\begin{pmatrix} 1 & 0 & 0 \\ 0 & \gamma & 0 \\ 0 & 0 & \gamma \end{pmatrix}\right)$,

$H_4 = C_G\left(\begin{pmatrix} \gamma & 0 & 0 \\ 0 & 1 & 0 \\ 0 & 0 & \gamma \end{pmatrix}\right)$, $H_5 = G$. Let $\theta \in \hat{T}_0^F$. Then it turns out that $K_T^G(\theta) = 2\,\mathrm{Ind}_{T_0^F}^{G^F}(\theta) - \mathrm{Ind}_{H_2^F}^{G^F}(R_{T_0}^{H_2}(\theta)) - \mathrm{Ind}_{H_3^F}^{G^F}(R_{T_0}^{H_3}(\theta))$

$- \mathrm{Ind}_{H_4^F}^{G^F}(R_{T_0}^{H_4}(\theta)) + R_{T_0}^G(\theta)$. It can be checked that

$\dim K_T^G(\theta) = (q^2-1)(q^3-1)(2q+1)$, whereas $|G^F|_{p'} = (q-1)(q^2-1)(q^3-1)$.

We now prove the main theorem.

<u>Theorem 7.13</u> ([41], p. 278). The class function $\chi_{G,T,\theta}$ on G^F is a virtual (i.e. generalized) character of G^F.

<u>Proof</u>. By induction on the semisimple rank of G we can assume that $\chi_{H,T,\theta}$ is a virtual character of H^F for any $H \in \mathcal{H}$, $H \neq G$. Thus it is sufficient to show that $K_T^G(\theta)$ is a virtual character of G^F. By (7.10) $K_T^G(\theta)$ has its support on $Z(G^F)V^F$. By Brauer's characterization of characters (see [26], 15.3; [66], p. 82, or [17], 40.8) it is sufficient to show that $K_T^G(\theta)$ is a virtual character on subgroups of the form $S \times R$ where S,R are subgroups of G^F consisting of semisimple elements and unipotent elements respectively. We can assume that $Z = Z(G^F) \subset S$. Then the support of $K_T^G(\theta)$ on SR is contained in ZR. So the restriction of $K_T^G(\theta)$ to SR is $\frac{|Z|}{|S|}\,\mathrm{Ind}_{ZR}^{SR}(K_T^G(\theta))$.

By Theorem 7.8, and the definition of $K_T^G(\theta)$, $K_T^G(\theta)$ is a virtual character on R. Then (7.12) shows that $|Z|(K_T^G(\theta)(u))$ is an integer and that it is divisible by $|C(u)^F|_{p'}$. Thus $\frac{|Z|(K_T^G(\theta)(u))}{|S|}$ is an integer. Now for any finite group Q, if O is the ring of integers in an algebraic number field which is a splitting field for Q, and f is an O-valued class function on Q, then $|Q|f$ belongs to the ring of virtual characters of Q with coefficients in O (see e.g., [66], Theorem 23') and thus the index of the O-module of virtual characters of Q with coefficients in O in the O-module of O-valued class functions on Q is a divisor of $|Q|$. Hence, since $|S|$, $|Z|$ are prime to p, $K_T^G(\theta)$ is a virtual character on R, and hence on ZR (here we are using (7.11)). Thus the restriction of $K_T^G(\theta)$ to SR is a virtual character and this proves the theorem.

Finally we show ([41], Theorem 3) that the functions $X_{T,G,\theta}$ and $Tr(\varepsilon_T\varepsilon_G R_T^G(\theta))$ on G^F coincide. We can make an induction assumption that $X_{G,T,\theta}$ and $\varepsilon_T\varepsilon_G\, Tr(R_T^G(\theta))$ coincide on subgroups of the form $C^0(s)^F$ where $s \notin Z$. Thus $X_{G,T\,\theta} - \varepsilon_G\varepsilon_T Tr(R_T^G(\theta))$ has its support on ZV^F, and furthermore $X_{G,T,\theta}(su) - \varepsilon_G\varepsilon_T Tr(su,R_T^G\theta)$ $= \theta(s)(\Phi_T^G(u) - \varepsilon_G\varepsilon_T Q_T^G(u))$ $(s \in Z,\ u \in V^F)$. Hence we have

$$\frac{1}{|G^F|}\sum_{\substack{s\in Z\\ u\in V^F}} |X_{G,T,\theta}(su) - \varepsilon_G\varepsilon_T Tr(su,R_T^G(\theta))|^2$$

$$= \frac{|Z|}{|G^F|} \sum_u |\Phi_T^G(u) - \varepsilon_G \varepsilon_T Q_T^G(u)|^2 \leq 4 \frac{|Z|\,|W(T)^F|}{|T^F|},$$

using (7.6) and (6.16). But the left hand side is an integer since $X_{G,T,\theta} - \varepsilon_G\varepsilon_T \mathrm{Tr}(R_T^G(\theta))$ is a virtual character of G^F. By our assumption on p being large, the right hand side must then be zero. (See (2.10), which describes how $|T^F|$ can be computed; $|W(T)^F|$ is independent of q.)

Remarks. 1. Kazhdan's theorem can be interpreted, for large p, as providing an alternative approach to the work of Lusztig-Deligne at the level of characters by constructing families of virtual characters corresponding to tori of G^F. We note, however, that étale cohomology is used here also in the proof of the crucial result (7.8). We refer the reader to [48], Remark 2.14, where the connection between the $K_T^G(\theta)$ and the $R_T^G(\theta)$ is made and $K_T^G(\theta)$ is interpreted as the alternating trace of G^F on the cohomology of a certain variety.

2. There are certain results in the literature about the values of unipotent characters (these will be defined in Chapter VIII) at semisimple elements of G^F. See [15], [14], and the references given there.

CHAPTER VIII. CLASSIFICATION OF REPRESENTATIONS.

In this chapter we turn to the classification of all the irreducible representations of a group of the form G^F. We will describe the classification of the representations of $GL(n,q)$ and $U(n,q)$ [50] and of certain versions of the other classical groups ([47], [48]). In the case of $GL(n,q)$ there is a recursive method of obtaining the Green functions and hence in principle we have the character table. By work of Hotta and Springer [37], along with the classification of the representations of $U(n,q)$, the same is true for $U(n,q)$ for large p. However in the case of Lusztig's classification of the representations of the other classical groups, we have only the dimensions but not the characters of the representations.

In some sense the so-called unipotent representations of G^F are the most important ones to obtain. Lusztig has classified the unipotent representations in all cases, ie. also in the case of the exceptional groups (see [48], [49]), and given their dimensions.

Definition. An irreducible representation of G^F is unipotent if it occurs as a constituent of $R_T^G(1)$ for some F-stable maximal torus T of G.

We note that the irreducible constituents of $\mathrm{Ind}_{B_0{}^F}^{G^F}(1) = R_{T_0}^G(1)$ are unipotent; in particular the representations 1 and St_G are unipotent. But there may be other unipotent representations of G^F (eg. θ_{10} of [76], which is in fact a cuspidal unipotent representation, and therefore cannot occur in $R^G_{T_0}(1)$ by (6.25)). The first proposition gives 1 as a linear combination of the $R^G_T(1)$.

Notation. $\sum_{(T)}$ stands for the sum over a set of representatives of G^F-conjugacy classes of F-stable maximal tori of G^F.

Proposition 8.1 ([48], 7.14; [48], 2.7). We have

$$1 = \sum_{(T)} \frac{1}{|W(T)^F|} R^G_T(1).$$

Proof. It is sufficient to show that $(R^G_T(1),1)_{G^F} = 1$ for any T since $R^G_T(1)$, $R^G_{T'}(\theta)$ are disjoint if $\theta \in \tilde{T}'^F$, $\theta \neq 1$ by (6.12) and $(R^G_T(1), R^G_T(1)) = |W(T)^F|$ by (6.14).

As before let $\tilde{X}$ denote the variety from which $R^G_T(1)$ is constructed. We have

$$\begin{aligned}(R^G_T(1),1) &= \sum_i (-1)^i \dim(H^i_c(\tilde{X})_1)^{G^F} \\ &= \sum_i (-1)^i \frac{1}{|T^F|} \sum_{t\in T^F} \mathrm{Tr}(t,H^i(\tilde{X})^{G^F}) \\ &= \sum_i (-1)^i \frac{1}{|T^F|} \sum_{t\in T^F} \mathrm{Tr}(t,H^i(G^F/\tilde{X})) \text{ by } (5.10)\end{aligned}$$

Now the map $g \to g^{-1}(Fg)$ gives rise to an isomorphism $G^F/\tilde{X} \tilde{\to} U$ (where $\tilde{X} = L^{-1}U$) and the action $g \to gt$ of T^F on $\tilde{X}$ corresponds to the action $u \to t^{-1}ut$ of T^F on U. But then this is just conjugation and the action of T^F can be extended to an action of T on U. Then the action of T on $H_c^i(U)$ is trivial by (6.5). This means $\mathcal{L}(t,U)$ is a constant for all t. Choose t_0 to be a regular element of T(ie. $C^\circ(t_0) = T$) so that t_0 acts fixed point freely on U. Then $\mathcal{L}(t_0,U) = \mathcal{L}(1,U^{t_0}) = 1$, using (6.3). So $\mathcal{L}(t,U) = 1$ for all t, and $(R_T^G(1),1) = \frac{1}{|T^F|} \sum_{t\in T^F} \mathcal{L}(t,U) = 1.$

<u>Corollary</u> 8.2. For any unipotent element u,

$$\sum_{(T)} \frac{1}{|W(T)^F|} Q_T^G(u) = 1.$$

<u>Remark</u> 8.3. It can also be shown ([18], 7.14) that

$$\sum_{(T)} \frac{T}{|W(T)^F|} R_T^G(1) = St_G.$$

By the classification of tori (2.8) each F-stable maximal torus T corresponds to an F-conjugacy class of $W = W(T_0)$. We write R_w for $R_T^G(1)$ if T corresponds to $w \in W$. By (2.9) we have $|W(T)^F| = |C'(w)|$. Thus we can write (8.1) as

$$(8.4) \qquad \frac{1}{|W|} \sum_{w\in W} R_w = 1.$$

Let $\mathcal{E}(W)$ denote the set of isomorphism classes of irreducible representations of W. Then F acts on $\mathcal{E}(W)$. Let $E \in \mathcal{E}(W)^F$, and we can ask whether, in analogy with (8.3) and (8.4), $\frac{1}{|W|} \sum_{w\in W} Tr(w,E)R_w$ is a unipotent representation of G^F. This is true if $G^F = GL(n,q)$ or $U(n,q)$ (with a slight modification, ie. "twisting by w_0") and furthermore in these cases all unipotent representations are obtained this way. But it is not true in general.

At this stage we need to generalize the construction of $R_T^G(\theta)$ to the case of an F-stable reductive subgroup L of G which is a Levi subgroup of a parabolic subgroup, and a character of L^F (see [44]).

Let $P = LV$ be the Levi decompotition of a parabolic subgroup P of G such that L is F-stable but P is not necessarily F-stable. We call L a <u>regular</u> subgroup of G (see [47], 7.2).

<u>Definition</u> 8.5. Let L be a regular subgroup of G. Then $S_{L\subset P,G} = \{g\in G \mid g^{-1}(Fg) \in V\}$.

We also denote the scheme $S_{L\subset P,G}$ by S when the context is clear. We have an action of $G^F \times L^F$ on S as $(g_0,\ell):g \to g_0 g \ell^{-1}$, and thus on $H_c^i(S)$. Let Π be any L^F-module (over $\bar{Q}_\ell$). Then $H_c^i(S) \otimes \Pi$ can be regarded as a $G^F \times L^F$ module by letting G^F act trivially on Π. Thus $(H_c^i(S) \otimes \Pi)^{L^F}$ is G^F-stable.

<u>Definition</u> ([44], p. 203) $R_L^G(\Pi) = \sum_i (-1)^i \, (H_c^i(S) \otimes \Pi)^{L^F}$.

$R_L^G(\Pi)$ is a virtual representation of G^F and we get a map $\Pi \to R_L^G(\Pi)$ of $\mathcal{R}(L^F)$ into $\mathcal{R}(G^F)$. Note that if $L = T$, a torus, and $\Pi = \theta \in \hat{T}^F$ we recover $R_T^G(\theta)$, for the action of T^F on S considered here is the inverse of the action on $\tilde{X}$ considered earlier.

Proposition 8.6. ([44], p. 204) Let T be an F-stable maximal torus and suppose $T \subset L$ where L is a regular subgroup. Let $\theta \in \hat{T}^F$. Then $R_L^G R_T^L(\theta) = R_T^G(\theta)$.

Proof. Choose a Borel subgroup B of G such that $T \subset B \subset P$, where $P = LV$. Let $B_1 = B \cap L$, $B_1 = TV_1$, $B = TU$, where V_1, U are unipotent. Then we have $U = V_1 V$, a semidirect product. Also B_1 is a Borel subgroup of L. We construct an isomorphism

$$L^F / S_{L\subset P,G} \times S_{T\subset B_1,L} \to S_{T\subset B,G} \text{ by } (g, g') \to gg'$$

where L^F acts on $S_{L\subset P,G} \times S_{T\subset B,L}$ by $(g, g') \to (g\ell^{-1}, \ell g')$

$(\ell \in L^F)$. It is easy to check that if $g \in S_{L\subset P,G}$, $g' \in S_{T\subset B_1,L}$ then $gg' \in S_{T\subset B,G}$. Conversely if $g'' \in S_{T\subset B,G}$ let $g''^{-1}(Fg'') = u_1 u_2$ with $u_1 \in V_1$, $u_2 \in V$. Write

$u_1 = g'^{-1}(Fg')$ where $g' \in L$ by Lang's Theorem (2.4). Let $g = g''g'^{-1}$. Then $g \in S_{L\subset P,G}$. Now if instead of g' we choose $g_0 \in L$ with $g_0^{-1} F(g_0) = u_1$, we have $g_0 = \ell g'$ where $\ell \in L^F$ and g is replaced by $g\ell^{-1}$. Thus we have the required isomorphism.

By the Künneth Formula (5.9) and by (5.10) we now have, for any non-negative integer r,

$$H_c^r(S_{T\subset B,G}) \simeq \bigoplus_{i+j=r} (H_c^i(S_{L\subset P,G}) \otimes H_c^j(S_{T\subset B_1,L}))^{L^F}.$$

This isomorphism is compatible with the action of $G^F \times T^F$.

Let $\theta \in \hat{T}^F$, and consider

$$H_c^r(S_{T\subset B,G})_{\theta^{-1}} \cong \bigoplus_{i+j=r} (H_c^i(S_{L\subset P,G}) \otimes H_c^j(S_{T\subset B_1,L})_{\theta^{-1}})^{L^F}.$$

Taking alternating sums over r we get

$$R_T^G(\theta) = \sum_r (-1)^r H_c^r(S_{T\subset B,G})_{\theta^{-1}}$$

$$= \sum_{i,j} (-1)^{i+j} (H_c^i(S_{L\subset P,G}) \otimes H_c^j(S_{T\subset B_1,L})_{\theta^{-1}})^{L^F}$$

$$= \sum_j (-1)^j R_L^G (H_c^j (S_{T\subset B_1,L})_{\theta^{-1}})$$

$$= R_L^G \; R_T^L (\theta), \text{ as required.}$$

Remark. It can be shown ([44], p. 208) that dim $\varepsilon_G \varepsilon_L R_L^G(\pi)$ $= |G^F|_{p'} \cdot |L^F|_{p'}^{-1}$ dim π.

Examples.

1. $G = GL_3$, $\tilde{F}$ the twisted Frobenius morphism so that $G^{\tilde{F}} = U(3,q)$. Let P be the parabolic subgroup of G consisting of matrices of the form $\begin{pmatrix} * & * & * \\ 0 & \multicolumn{2}{c}{\boxed{*}} \\ 0 & & \end{pmatrix}$, with the obvious Levi subgroup

$L = \left\{ \begin{pmatrix} * & 0 & 0 \\ 0 & \multicolumn{2}{c}{\boxed{*}} \\ 0 & & \end{pmatrix} \right\}$. Then L, but not P, is $\tilde{F}$-stable and $L^{\tilde{F}} \cong U(1,q) \times U(2,q)$. We have two tori T_0, T_1 in L such that $|T_0^{\tilde{F}}| = (q+1)(q^2-1)$, $|T_1^{\tilde{F}}| = (q+1)^3$. A typical element of $L^{\tilde{F}}$

can be taken as $\begin{pmatrix} \eta^a & 0\ 0 \\ 0 & A \\ 0 & \end{pmatrix}$ where η is a primitive (q+1)st root of unity in K and $A \in U(2,q)$. Let $\phi_{\ell,m}$ be the character of $\tilde{L}^F$ which takes this element to $(\det A)^M \eta^{\ell a}$ (interpreted as an element of $\bar{\mathbb{Q}}_\ell$ by choosing an embedding of K^* in $\bar{\mathbb{Q}}_\ell^*$) where ℓ, m are integers such that $1 \leq \ell, m \leq q+1$. We obtain a family $\{R_L^G(\phi_{\ell,m})\}$ of virtual representations of $\tilde{G}^F$ of dimension q^2-q+1 which are irreducible if ℓ, m are distinct. We also have a family $\{R_L^G(St_L \cdot \phi_{\ell,m})\}$ of virtual representations of $\tilde{G}^F$ of dimension $q(q^2-q+1)$ which are irreducible if $\ell \neq m$. In fact, we have $R_L^G(\phi)$ $= \frac{1}{2}[R_{T_0}^G(\phi) + R_{T_1}^G(\phi)]$, $R_L^G(St_L \cdot \phi) = \frac{1}{2}[R_{T_0}^G(\phi) - R_{T_1}^G(\phi)]$, writing $\phi = \phi_{\ell,m}$. [Here we have restricted ϕ to $\tilde{T}_0^F$ and $\tilde{T}_1^F$.]

We have similar representations for $GL(3,q)$, but they are less interesting since L and P are both stable under the standard Frobenius morphism.

2. $G = GL_4$, F the standard Frobenius so that $G^F = GL(4,q)$. Let $L = \left\{\begin{pmatrix} A & 0 \\ 0 & B \end{pmatrix}\right\} \simeq GL_2 \times GL_2$ and consider the element $w = \begin{pmatrix} 0 & I \\ I & 0 \end{pmatrix}$ in $N(L)$. If $w = a(Fa)^{-1}$, $a \in G$, then $(a^{-1}La)^F \subset G^F$ is isomorphic to $GL(2,q^2)$, and $(a^{-1}La)^F$ is not contained in a

parabolic subgroup of G^F as a Levi subgroup. If θ is a linear character of $(a^{-1}La)^F$ we have a corresponding virtual representation $R^G_{a^{-1}La}(\theta)$ of G^F of dimension $(q-1)(q^3-1)$ which is irreducible if it is not fixed by a non-trivial element of $N(a^{-1}La)^F/(a^{-1}La)^F$. Thus we get a family of virtual representations of dimension $(q-1)(q^3-1)$ of $GL(4,q)$ and by using $St_{a^{-1}La} \cdot \theta$ we get another family of virtual representations of dimension $q^2(q-1)(q^3-1)$. If we use the twisted Frobenius morphism $\tilde{F}$ in G, so that $G^{\tilde{F}} \cong U(4,q)$ and the same subgroup L we get families of virtual representations of $U(4,q)$ of dimensions $(q+1)(q^3+1)$, $q^2(q+1)(q^3+1)$ (see [57], where these are denoted by $\chi_3(k)$ and $\chi_4(k)$).

3. Example 1 can be generalized as follows. Let $G = GL_n$, $\tilde{F}$ the twisted Frobenius morphism, $G^{\tilde{F}} = U(n,q)$. Let

$$L = \left\{ \begin{pmatrix} * & 0 & 0 \\ 0 & \boxed{*} & \\ 0 & & \end{pmatrix} \right\} \cong GL_1 \times GL_{n-1}, \text{ so that } L^{\tilde{F}} \cong U_1(n,q) \times U_{n-1}(n,q).$$

For any character θ of $L^{\tilde{F}}$ we can construct $R^G_L(\theta)$.

The virtual representation $R^G_L(1)$ is the sum of the trivial representation and a virtual representation Ψ which is (up to sign) irreducible. The representation Ψ (when n is even) was constructed by Tate and Thompson [85] and historically is the first example of a representation of a finite classical group constructed using étale cohomology. (See [36] for more details, including the character of Ψ.)

Representations of GL(n,q) and U(n,q).

We now describe the classification of representations of GL(n,q) and U(n,q). The unipotent representations of GL(n,q) were constructed by Steinberg [78]. The classification of all the representations of GL(n,q) was given by Green [30]. The classification of the representations of U(n,q) was given by the author and Lusztig in [50]. This method treats the cases of GL(n,q) and U(n,q) simultaneously and will be described below. A long-standing conjecture of Ennola [23] stated that the Green functions of U(n,q) could be obtained from the Green functions of GL(n,q) by changing q to $-q$. This has been proved to be true for large p (see [37]). Thus in principle we can construct the character table of U(n,q) for large p.

We now let $G = GL_n$ and denote by F either the standard (Case 1) or the twisted (Case 2) Frobenius morphism, so that G^F is either GL(n,q) in Case 1 or U(n,q) in Case 2. We recall that B_0 is the group of upper triangular matrices, T_0 the group of diagonal matrices, and $W = W(T_0)$ is isomorphic to S_n, the symmetric group of degree n. T_0 is F-stable in both cases but B_0 is F-stable only in Case 1.

Let w_0 be the element of maximal length in W (see e.g. [8], p. 43). (We can take $\dot{w}_0 = \begin{pmatrix} & & 1 \\ & \cdot^{\cdot^{\cdot}} & \\ 1 & & \end{pmatrix}$ in $N(T_0)$.)

Let $e = 0$ in Case 1 and $e = 1$ in Case 2. Then we have $w_0^{-e} w w_0^{e} = Fw$ for each $w \in W$. Let $\Delta \subset \Phi^+$ be the set of simple roots which determine B_0. For each $J \subset \Delta$, we have a subgroup W_J of W generated by reflections in the hyperplanes orthogonal to the roots in J in the vector space $V = X(T_0) \otimes_{\mathbb{Z}} \mathbb{R}$. The parabolic subgroups $P_J = B_0 W_J B_0$ are called standard parabolic subgroups containing B_0, and we have Levi decompositions $P_J = L_J V_J$ where L_J contains T_0 (see [39], p. 183). In our case each L_J is a product of GL_k's for $k \leq n$. We have, for each J, $\dot{w}_0^{-e} L_J \dot{w}_0 = FL_J$. If $\dot{w}_0^e = y^{-1}(Fy)$ $(y \in G)$ then $L = yL_J y^{-1}$ is F-stable. (Note that we can take $y = 1$ in Case 1; in Case 2, L_J^F is a direct product of finite unitary groups and is not necessarily contained in a parabolic subgroup of G^F.)

Proposition 8.7. For each $J \subset \Delta$, $\frac{1}{|W_J|} \sum_{w \in W_J} R_{ww_0}$ is a virtual representation of G^F.

(Here the notation R_w is as in (8.4).)

Proof. We note that $\dot{w}_0^{-e} B_0 \dot{w}_0^e = FB_0$ and hence $yB_0 y^{-1}$ is F-stable. So the canonical torus (ie. the analogue of T_0) in L can be taken to be $T_1 = yT_0 y^{-1}$. Also $N_L(T_1) = yN_{L_J}(T_0)y^{-1}$ and we can take as a set of representatives in $N_L(T_1)$ for $W_L = N_L(T_1)/T_1$ the set $\{y\dot{w}y^{-1} \mid w \in W_J\}$. Suppose $a^{-1}T_1 a$ $(a \in L)$ is F-stable, ie. $a^{-1}yT_0 y^{-1}a$ is F-stable. Then $y^{-1}a(Fa)^{-1}(Fy) \in N(T_0)$ and since we can write $a(Fa)^{-1} = y\dot{w}y^{-1}$

for some $w \in W_J$, we have $y^{-1}(aFa)^{-1}(Fy) = \dot{w}\dot{w}_0^e$ for some $w \in W_J$. Now by (8.4) we have that $\frac{1}{|W_L|} \sum_{x \in W_L} R^L_{T_x}(1) = 1_L$, where T_x is a maximal torus of L corresponding to the element x of W_L. Using (8.6) we see that

$\frac{1}{|W_L|} \sum_{x \in W_L} R^G_{T_x}(1)$ is a virtual representation of G^F. But the above remarks show that this virtual representation can be written as $\frac{1}{|W_J|} \sum_{w \in W_J} R_{ww_0^e}$.

<u>Theorem 8.8</u>. Let $E \in \mathcal{E}(W)$. Then $\frac{1}{|W|} \sum_{w \in W} Tr(ww_0^e, W)R_w$ is a unipotent representation (up to sign) of G^F and all unipotent representations of G^F arise this way.

<u>Proof</u>. By a theorem of Frobenius [27] every irreducible representation E of W is an integral linear combination of representations of the form $Ind^W_{W_J}(1)$, for some $J \subset \Delta$.

Let $E = Ind^W_{W_J}(1)$, for some J. Then

$$\frac{1}{|W|} \sum_{w \in W} Tr(ww_0^e, E)R_w = \frac{1}{|W|} \sum_{w \in W} Tr(w, E)R_{ww_0^e}$$

$$= \frac{1}{|W_J|} \sum_{w \in W_J} R_{ww_0^e}.$$

By (8.7), the left hand side is a virtual representation and this now holds for any E. We have to show that it is irreducible up to sign. By the orthogonality relations (6.14) and by (2.9) we have that $(R_w, R_w) = |C'(w)|$. Using the fact that $w_0^{-e} w w_0^e = Fw$ for each $w \in W$, we see that $C'(ww_0^e) = C(w)$. Using the orthogonality relations for the characters of W we then see that $\frac{1}{|W|} \sum_{w \in W} Tr(w,E)\, R_{ww_0}$ has norm 1, as required.

We now have as many unipotent representations as the number of conjugacy classes of W, which in this case is also equal to the number of F-conjugacy classes of W and hence to the number of distinct R_w. Also by inverting the equations we see that the R_w can be expressed in terms of these unipotent representations. Thus we have constructdd all the unipotent representations of G^F.

Remark. The number of unipotent representations is also equal to the number of unipotent conjugacy classes of G^F, each being equal to the number of partitions of n.

We will now describe how the representations of $GL(n,q)$ and $U(n,q)$ are constructed. For details and proofs see [50]. Choose a set of representatives $\{L\}$ for the G^F-conjugacy classes of centralizers of F-stable semisimple elements of G. Consider linear characters θ of L^F which are trivial on $D(L)^F$ where $D(L)$ is the derived group of L. We say θ is regular if it is not fixed by any non-trivial element of $N(L)^F/L^F$. Choose a set $\mathcal{S}$ of representatives for the $N(L)^F/L^F$-orbits of such regular characters of L^F and let $\Pi \in \mathcal{E}(L)$ be of the form $\Pi = \theta\phi$ where $\theta \in \mathcal{S}$ and ϕ is a unipotent representation of L^F. Then, as (L,π) varies over all such pairs, the virtual representations $R_L^G(\pi)$ vary over all irreducible representations of G^F, up to sign.

Remark. We have shown that the unipotent representations of G^F (up to sign) are given by $\frac{1}{|W|}\sum_{w\in W}\chi(w)\,R_{ww_0}$, where χ is an irreducible character of $W \cong S_n$. We discuss now when they are cuspidal.

Case 1. $GL(n,q)$. Then all these representations are contained in $\mathrm{Ind}_{B_0^F}^{G^F}(1)$, and so by (6.25) they are not cuspidal.

Case 2. $U(n,q)$. (See [47], §9.) Suppose such a representation is cuspidal. By (6.25), whenever ww_0 corresponds to a non-minisotropic torus $\chi(w) = 0$. This means that whenever w has an eigenvalue -1 in the natural representation of W on $X(T_0)\otimes\mathbb{R}$, $\chi(w) = 0$. Thus χ vanishes on all elements of even order and by a theorem of Brauer and Nesbilt (see eg. [17], §86), $\dim\chi$ must be divisible by the highest power of 2 dividing $|G^F|$. Now we have a formula for the dimension of χ : $\dim\chi = \frac{n!}{\prod_{i,j} h_{ij}}$, where h_{ij} is the hook length of the square (i,j) in the Young diagram corresponding to χ. Thus all the h_{ij} must be odd, and the only possible such Young diagram corresponds to the partition $n = 1+2+\ldots+k$, for some k. Thus $n = \frac{k(k+1)}{2}$. Conversely if n is of this form and χ corresponds to the partition of n as above then we get a unipotent cuspidal representation of G^F which corresponds to χ. This is the unique unipotent cuspidal representation of G^F.

Example. $U(3,q)$ has a unipotent cuspidal representation of dimension q^2-q. In this case $\mathrm{Ind}_{B_0{}^F}^{G^F}(1)$ has two constituents whereas W has three irreducible representations, and this accounts for the extra unipotent representation.

The description of the representations of $GL(n,q)$ and $U(n,q)$ that we have given does not hold in general and one reason why it works here is that $G = GL_n$ is a "self-dual" group. We now introduce the idea of a "dual group" of G. (This idea is also related to Langlands's notion of an L-group.)

Dual group ([47], §7).

Let G,F be as usual and T any F-stable maximal torus of G. Recall (Chapter VI) that $X = X(T) = \mathrm{Hom}(T,K^*)$ an $Y = Y(T) = \mathrm{Hom}(K^*,T)$ are $\mathbb{Z}$-duals of each other. We can think of the set of roots Φ as a subset of X. We have a subset Φ^V of Y which defines a "dual root system" to Φ (see [8], p. 277) and the elements of Φ^V are called coroots. The data $\{X,Y,\Phi,\Phi^V\}$ determines G up to isomorphism (see [61], Expose XXV). Now F acts on X and Y. The group G^F is determined if we know the actions of F and of $q^{-1}F$ (which has finite order) on Y. Now we just reverse Y and X and replace the root system by its dual; ie. we consider the data $\{Y,X,\Phi^V,\Phi\}$ which determine a connected reductive group G^*, the dual of G.

Then G^* is also defined over $\mathbb{F}_q$ and has an F-stable maximal torus T^* which is dual to T in the sense that $X(T^*) = Y(T)$, $Y(T^*) = X(T)$. The Weyl groups $W(T)$ and $W(T^*)$ can be identified. There is a natural correspondence between characters of T^F and elements of T^{*F}, since $\hat{T}^F \simeq X(T)/(F-1)X(T)$ and $T^{*F} \simeq Y(T^*)/(F-1)Y(T^*)$ (see (6.9); a similar exact sequence $0 \to X(T) \xrightarrow{F-1} X(T) \longrightarrow \hat{T}^F \to 0$ is given by taking the map $X(T) \to \hat{T}^F$ to be restriction of characters to T^F, composed with a fixed isomorphism of K^* into $\bar{Q}_\ell^*$). Thus, if $\theta \in \hat{T}^F$, we have an element $s \in T^{*F}$ which is well-defineed up to conjugacy by an element of $N(T^*)^F$. Instead of $R_T^G(\theta)$ we can talk of $R_{T^*}^G(s)$ where s is an element of G^{*F}. Then the Strong Orthogonality Theorem (6.13) can be rephrased as follows.

<u>Theorem 8.9</u> ([12], 5.21 and [47], 7.5.2). The virtual representations $R_{T_1^*}^G(s_1)$, $R_{T_2^*}^G(s_2)$ are disjoint unless s_1 and s_2 are G^{*F}-conjugate.

We also say the finite groups G^F, G^{*F} are duals of one another There is a bijection between G^F-conjugacy classes of regular subgroups of G and G^{*F}-conjugacy classes of regular subgroups of G^{*F}.

Examples of dual groups.

G	G^*
SL_n	PGL_n
GL_n	GL_n
Sp_{2n}	SO_{2n+1}
$SO^{\pm}_{2n}$	$SO^{\pm}_{2n}$
CSp_{2n}	G^{o}_{2n+1}
$CO^{\pm,o}$	$G^{\pm,o}_{2n}$

G^F	G^{*F}
$SL(n,q)$	$PGL(n,q)$
$GL(n,q)$	$GL(n,q)$
$U(n,q)$	$U(n,q)$
$Sp(2n,q)$	$SO(2n+1,q)$
$SO^{\pm}(2n,q)$	$SO^{\pm}(2n,q)$

Here CSp_{2n} denotes the group of symplectic similitudes and $CO^{\pm,o}_{2n}$ is the identity component of the group of similitudes of a quadratic form over $\mathbb{F}_q$ which may be split (+) or non-split(-). G^{o}_{2n+1}, $G^{\pm,o}_{2n}$ are the special Clifford groups of a quadratic form over $\mathbb{F}_q$ in dimensions 2n+1, 2n. There exist homomorphisms $G^{o}_{2n+1} \to SO_{2n+1}$, $G^{\pm,o} \to SO^{\pm}_{2n}$ which give rise to surjective homomorphisms of the finite groups $(G^{o}_{2n+1})^F \to (SO_{2n+1})^F$, $(G^{\pm,o}_{2n+1})^F \to (SO^{\pm}_{2n})^F$ with central kernels of order q-1 (see [47], p. 154).

The next section will be devoted to describing the centralizer algebras of representations of G^F induced from certain representations of parabolic subgroups. These algebras are also called Hecke algebras in the literature.

Centralizer algebras.

References for this section are ([14], §3), [16], ([45], §5), ([47], §5), and ([8], pp. 54-55).

We recall that as a result of the Harish-Chandra Theory described in Chapter IV, one of our problems is to decompose representations of G^F of the form $\mathrm{Ind}_{P^F}^{G^F}(\tilde{\rho})$ where $\tilde{\rho}$ is the lift to P^F of a cuspidal representation ρ of L^F. Every irreducible representation of G^F is a constituent of such a representation.

In particular, consider $\mathrm{Ind}_{B_0{}^F}^{G^F}(1)$ and more generally $\mathrm{Ind}_{B_0^F}^{G^F}(\tilde{\lambda})$ where $\lambda \in \hat{T}_0^F$. These representations give rise to the principal series of G^F. The former, ie. $\mathrm{Ind}_{B_0^F}^{G^F}(1)$ has been studied by Curtis, Iwahori and Kilmoyer [16] and the latter, ie. $\mathrm{Ind}_{B_0{}^F}^{G^F}(\tilde{\lambda})$ by Kilmoyer [42] in the case when G is of adjoint type and by Kilmoyer and Howlett [38] in general (in fact, when G^F is a finite group with a split BN-pair). We will begin with describing the centralizer algebra of $\mathrm{Ind}_{B_0{}^F}^{G^F}(1)$.

Let (W,S) be a finite Coxeter group (see [8]). This means that W is a finite group with distinguished generators

$\{w_i\}$ $(w_i \in S)$ and defining relations $(w_iw_j)^{m_{ij}} = 1$ where $m_{ii} = 1$ for all i. If $w \in W$, $\ell(w)$ is the minimal number of generators appearing among all possible expressions of w as a product of distinguished generators.

Let $R = \mathbb{Q}[X_1, X_2, \ldots, X_n]$ denote the polynomial ring over $\mathbb{Q}$ generated by indeterminates $\{X_i\}$ corresponding to the generators w_i in S, such that $X_i = X_j$ whenever w_i and w_j are conjugate in W. For example, if W is a Weyl group of an indecomposable root system, then R will be a polynomial ring in one or two generators. We then define a <u>generic algebra</u> A_W over R with basis $\{T_w\}$ $(w \in W)$, where $T_1 = 1$ and multiplication is defined by

$$(8.10) \qquad T_{w_i}T_w = \begin{cases} T_{w_iw} & \text{if } \ell(w_iw) \geq \ell(w) \\ X_iT_{w_iw} + (X_i - 1)T_w & \text{if } \ell(w_iw) < \ell(w) \end{cases}$$

It can be shown (see [16], 1.8) that A_W has a presentation with relations

$$T_{w_i}^2 = X_i1 + (X_i - 1)T_{w_i},$$

$$\underbrace{T_{w_i}T_{w_j}T_{w_i}\cdots}_{m_{ij}} = \underbrace{T_{w_j}T_{w_i}T_{w_j}\cdots}_{m_{ij}} \qquad (i \neq j).$$

Now suppose we have a <u>specialization</u> of R, i.e., a homomorphism $f: R \to \mathbb{Q}$, under which $X_i \to q_i \in \mathbb{Q}$. Then we get an algebra A_f over $\mathbb{Q}$ by taking basis elements $\{T_{\bar{w}}\}$ $(w \in W)$ and multiplication defined by (8.10) with X_i replaced by q_i.

If we now consider a group of the form G^F, it turns out that the centralizer algebra of $\mathrm{Ind}_{B_0^F}^{G^F}(1)$ is isomorphic to an algebra A_f for a suitable specialization of R, with the Coxeter group being $W(T_0)^F$. This follows from a result of Iwahori and Matsumoto (see [16], 1.6) giving a presentation for the centralizer algebra of $\mathrm{Ind}_{B_0^F}^{G^F}(1)$. For example, if G is split over $\mathbb{F}_q$ (i.e., T_0 is split over $\mathbb{F}_q$) and F is the standard Frobenius morphism, then $f(X_i) = q$ for all i. If $G = GL_n$ and F is the twisted Frobenius so that $G^F = U(n,q)$, then we can take $R = \mathbb{Q}[X_1,X_2]$ and $f(X_1) = q$, $f(X_2) = q^2$ if n is even, $f(X_1) = q^2$, $f(X_2) = q^3$ if n is odd. Similarly Kilmoyer showed in [42] that if G is adjoint then for any $\lambda \in \hat{T}_0^F$, the centralizer algebra of $\mathrm{Ind}_{B_0^F}^{G^F}(\tilde{\lambda})$ is also of the form A_f, for a suitable generic algebra A_W.

Let $\mathfrak{F}$ be the quotient field of R, $\bar{\mathfrak{F}}$ an algebraic closure of $\mathfrak{F}$, and $\bar{R}$ the integral closure of R in $\bar{\mathfrak{F}}$. Consider the $\bar{\mathfrak{F}}$-algebra $A_W \otimes_R \bar{\mathfrak{F}}$ and let χ be an irreducible character of $A_W \otimes_R \bar{\mathfrak{F}}$. Then $\chi(T_w) \in \bar{R}$ for each $w \in W$ and if we use the specialization f mentioned above we get an irreducible character of the algebra $A_f \otimes_{\mathbb{Q}} \mathbb{C}$ which corresponds to a constituent of $\mathrm{Ind}_{B_0^F}^{G^F}(1)$. Moreover, there is an element $d_\chi \in \bar{\mathfrak{F}}$ called the <u>generic degree</u> corresponding to χ given by

$$d_\chi = \frac{(\dim \chi) \sum_{w\in W} \operatorname{Ind} T_w}{\sum_{w\in W} (\operatorname{Ind} T_w)^{-1} \chi(T_w)\chi(T_{w^{-1}})}$$

where Ind is the homomorphism of A_W into R defined by $\operatorname{Ind}(T_{w_i}) = X_i$. The crucial point about the generic degree is that again under the specialization f, or rather under an extension of f to a homomorphism of $\bar{R}$ into C, d_χ becomes the dimension of the constituent ϕ of $\operatorname{Ind}_{B_0^F}^{G^F}(1)$ corresponding to χ. Thus, if we can get explicit expressions for the d_χ in terms of the X_i, we have a way of computing the dimensions of the principal series representations of G^F. Generic degrees have been computed for the indecomposable Weyl groups in all cases except in type F_4 with $R = \mathbb{Q}[X_1, X_2]$; even in this case most of the generic degrees are known. For classical groups this was done by Hoefsmit [35] and for the exceptional groups by Benson, Grove and Surowski [4] and by Benson [3].

Now we turn to more recent work of Lusztig ([45], [47], [48]) involving centralizer algebras of representations of G^F of the form $\operatorname{Ind}_{P^F}^{G^F}(\tilde{\rho})$ where ρ is a unipotent cuspidal representation of L^F. First we give an alternative description of the specializations of the generic algebra. Let (W,S) be a Coxeter group and let $\phi\colon S \to \overline{\mathbb{Q}}_\ell$ be a function such that $\phi(s) = \phi(s')$ whenever s, s' are conjugate in W. Let $\mathcal{H}(W,\phi)$ be the $\overline{\mathbb{Q}}_\ell$-algebra with basis $\{T_w\}$ $(w \in W)$ and multiplication defined by

$$T_w T_{w'} = T_{ww'}, \quad \text{if} \quad \ell(ww') = \ell(w) + \ell(w').$$

$$(T_s+1)(T_s-\phi(s)) = 0, \text{ for all } s \in S.$$

Next, we note that if ρ is a unipotent cuspidal representation of L^F then $\mathrm{Ind}_{P^F}^{G^F}(\tilde{\rho})$ contains only unipotent representations, and conversely if ρ is a unipotent cuspidal representation of G^F which corresponds to an F-stable parabolic subgroup P in the Harish-Chandra classification then ρ is a constituent of $\mathrm{Ind}_{P^F}^{G^F}(\tilde{\rho}_0)$ where ρ_0 is a unipotent cuspidal representation of L^F (see e.g. [13], 1.2). The following theorem follows from Lusztig's work.

Theorem 8.11. Let P be an F-stable parabolic subgroup of G and let $P = LV$ where L is F-stable. Let ρ_0 be a cuspidal unipotent representation of L^F. Then the centralizer algebra of $E = \mathrm{Ind}_{P^F}^{G^F}(\tilde{\rho}_0)$ is isomorphic to $\mathcal{H}(W_0,\phi)$ where (W_0,S) is a suitable Coxeter group and $\phi\colon\ S \to \bar{\mathbb{Q}}_\ell$ is a suitable function. The isomorphism is special in the sense that the basis elements T_w $(w \neq 1)$ of $\mathcal{H}(W_0,\phi)$ correspond to endomorphisms of E of trace zero.

We give a brief discussion of this theorem. Suppose G has a connected Dynkin graph Γ. Suppose we are given a unipotent non-cuspidal representation ρ of G^F. We can find a suitable F-stable parabolic subgroup P_1, where $P_1 = L_1V_1$ and L_1 is F-stable, such that ρ is a constituent of $\mathrm{Ind}_{P_1^F}^{G^F}(\tilde{\rho}_1)$ and ρ_1 is a unipotent cuspidal representation of L_1^F. Then L_1 also has a connected Dynkin graph Γ' and it can be shown that ρ_1 is uniquely determined by ρ. (In most cases, this follows from the fact that L_1^F has a unique unipotent cuspidal representation.) Moreover, Γ' is the unique F-stable subgraph of its type of Γ.

Conversely, suppose P is F-stable and L has a Dynkin graph Γ' which is an F-stable subgraph of Γ. Suppose ρ_0 is a unipotent cuspidal representation of L^F. Let $\bar{\Gamma}$ be a graph whose vertices are in bijection with the orbits of F on $\Gamma - \Gamma'$. For two such orbits γ, γ' we define

$$m_{\gamma,\gamma'} = \frac{2(|\Phi_{\Gamma'\cup\gamma\cup\gamma'}| - |\Phi_{\Gamma'}|)}{|\Phi_{\Gamma'\cup\gamma}| + |\Phi_{\Gamma'\cup\gamma'}| - 2|\Phi_{\Gamma'}|},$$

where $\Phi_{\Gamma'},\ldots$ denote the set of roots of a root system of type $\Gamma',\ldots$. Then $m_{\gamma,\gamma'} = 2, 3, 4$ or 6, and, as usual, we join the vertices of $\bar{\Gamma}$ corresponding to γ and γ' by 0, 1, 2 or 3 bonds in the four cases. The resulting graph is a graph of a Coxeter group W_0. A suitable function ϕ is then defined on the set S of simple reflections of W_0 and the centralizer algebra of E is shown to be isomorphic to $\mathcal{H}(W_0,\phi)$. For a case-by-case list of Γ, Γ', $\bar{\Gamma}$ and ϕ see [48], p. 35. A discussion of the above results is also on pp. 33-34, loc. cit.

In fact, the group W_0 is isomorphic to $(N(L)/L)^F$. It is not usually the case that this group turns out to be a Coxeter group (it is a section of the Weyl group $W(T_0)$ of G). Lusztig

([45], 5.9) proves that it is a Coxeter group if the Weyl group of L satisfies a certain assumption ([45], 5.7.1) on subsets of the set of simple reflections being stable under conjugation by the longest element. This assumption is satisfied in all the cases when L^F has a unipotent cuspidal representation. (For example, if L is of type A and G is of type D it is not satisfied, but untwisted groups of type A have no unipotent cuspidal representations. Another such example is when L is of type D_5 and G is of type E_6; again L^F has no unipotent cuspidal representations.)

Theorem 8.11 is proved when G is classical in [47], §5 (see especially 5.15). When G is exceptional it is proved in [45], §5; here it is assumed that the unipotent cuspidal representation ρ is contained in $R_T^L(1)$ where T is a "Coxeter torus." However, for E_6 and E_7 all unipotent cuspidal representations are of this form, whereas the other groups do not appear in the form of a subgroup L inside a bigger group G. Another assumption on ρ needed for the theorem is that it is extendible to a representation of $N(L)^F$. This is also true in all the cases; in most cases it follows because it is the only unipotent cuspidal representation. If this is not so, it follows from [45], Lemma 6.6.

<u>Representations of classical groups</u> ([47], [48]).

We have seen that in the case of $GL(n,q)$ and $U(n,q)$ the unipotent representations are parametrized by partitions of n.

Thus it would appear as a first guess that in the case of groups of types B, C, D (i.e., the classical groups), pairs of partitions have to be used since they parametrize the conjugacy classes of the Weyl group of type B. However, this does not work, and neither do we get formulas for the unipotent representations as in Theorem 8.8 (except for the trivial or Steinberg representations). In fact, the number of unipotent representations is not equal to the number of F-conjugacy classes of the Weyl group, in general. Lusztig has introduced certain combinatorial objects called "symbols" and shown that these parametrize the unipotent representations of finite classical groups. We will now describe this result.

The idea is this: a partition of n can either be thought of as a decreasing sequence of integers $\alpha_1 \geq \alpha_2 \geq \cdots \geq \alpha_m$ with $\alpha_1 + \alpha_2 + \cdots + \alpha_m = n$, or, taking $\lambda_i = \alpha_i - i + m$ $(1 \leq i \leq m)$ as an array of integers $\lambda_1 > \lambda_2 > \cdots > \lambda_m$. This is generalized as follows. A <u>symbol</u> is an array of the form

$$\Lambda = \begin{pmatrix} \lambda_1 > \lambda_2 > & & > \lambda_a \\ \mu_1 > \mu_2 > & & > \mu_b \end{pmatrix}$$

where the λ_i, μ_i are non-negative integers. We say that Λ is equivalent to $\begin{pmatrix} \lambda_1+1 > \lambda_2+1 > \cdots > \lambda_a+1 > 0 \\ \mu_1+1 > \mu_2+1 > \cdots > \mu_a+1 > 0 \end{pmatrix}$

or to $\begin{pmatrix} \mu_1 > \mu_2 > \cdots > \mu_b \\ \lambda_1 > \lambda_2 > \cdots > \lambda_a \end{pmatrix}$.

<u>Definition</u> (i) def $\Lambda = |a - b|$ is the <u>defect</u> of Λ.

(ii) $\text{rk}\,\Lambda = \sum_i \lambda_i + \sum_i \mu_i - [(\frac{a+b-1}{2})^2]$ is the <u>rank</u> of Λ.

The defect and rank are functions on the set of equivalence classes of symbols.

<u>Definition</u>. $\Phi_{n,d}$ is the set of equivalences classes of symbols of rank n and defect d.

<u>Definition</u>. If $\Lambda = \begin{pmatrix} \lambda_1 > \lambda_2 > \cdots > \lambda_a \\ \mu_1 > \mu_2 > \cdots > \mu_b \end{pmatrix}$, then

$$D_\Lambda(q) = \frac{\prod_{i<j} (q^{\lambda_i} - q^{\lambda_j}) \prod_{i<j} (q^{\mu_i} - q^{\mu_j}) \prod_{\substack{1\leq i\leq a \\ 1\leq j\leq b}} (q^{\lambda_i} + q^{\mu_j})}{\left(\prod_{i=1}^{a} (q^2-1)\cdots(q^{2\lambda_i}-1)\right)\left(\prod_{i=1}^{b} (q^2-1)\cdots(q^{2\mu_i}-1)\right) 2^c q^{\binom{a+b-2}{2}+\binom{a+b-4}{2}+\cdots}}$$

where $c = \begin{cases} [\frac{a+b-1}{2}], & \text{if } \{\lambda_1,\lambda_2,\ldots,\lambda_a\} \neq \{\mu_1,\mu_2,\ldots,\mu_b\} \\ a = b, & \text{otherwise.} \end{cases}$

<u>Theorem 8.12</u> ([47], 8.2). There exists a bijection between the unipotent representations of the groups

(i) $Sp(2n,q)$ and (ii) $SO(2n+1,q)$ and equivalence classes of symbols of rank n and odd defect

(iii) $SO^+(2n,q)$ and equivalence classes of symbols of rank n and defect $\equiv 0 \pmod 4$,

(iv) $SO^-(2n,q)$ and equivalence classes of symbols of rank n and defect $\equiv 2 \pmod 4$.

If the correspondence is $\{\Lambda\} \to \rho_\Lambda$ where Λ is a symbol and ρ_Λ is a unipotent representation of the group in question, ρ_Λ is cuspidal if and only if $\text{rk}\,\Lambda = n = [(\frac{\text{def}\,\Lambda}{2})^2]$, and there is at most one such representation. Cuspidal representations occur precisely when $n = k^2 + k$ for some k in Cases (i) and (ii), $n = k^2$ for some even k in Case (iii), and $n = k^2$ for some odd k in Case (iv). Finally we have

(8.13) $$\dim \rho_\Lambda = D_\lambda(q)\,|G^F|_{p'}\ ,$$

where G^F is a group as in (i), (ii), (iii) or (iv).

We will return to this theorem after stating another theorem (8.14), which brings out the importance of knowing the unipotent representations. As before we let G^* denote the dual group of G. We can write $\mathcal{E}(G) = \coprod_{(s)} \mathcal{E}(G,(s))$ where s runs over a set of representatives of G^{*F}-conjugacy classes of semisimple elements of G^{*F}, and

$$\mathcal{E}(G,(s)) = \{\rho \in (G) \mid (\rho, R^G_{T^*}(s)) \neq 0 \text{ for some F-stable maximal torus } T^* \subset G^*,\ s \in T^{*F}\}.$$

Note that $\mathcal{E}(G,(1))$ is the set of unipotent representations of G^F.

Theorem 8.14 ([47], 8.2). Let G be one of the following groups.

(a) $G = Sp_{2n}$, q even, $n \geq 1$; $G^* = SO_{2n+1}$.

(b) $G = SO^{\pm}_{2n}$, q even, $n \geq 2$; $G^* = SO_{2n}$.

(c) $G = SO_{2n+1}$, q odd, $n \geq 1$; $G^* = Sp_{2n}$.

(d) $G = CSp_{2n}$, q odd, $n \geq 1$; $G^* = G^0_{2n+1}$

(e) $G = CO^{\pm,0}_{2n}$, q odd, $n \geq 2$; $G^* = G^{\pm,0}_{2n}$.

Then there is a bijection, for each $s \in G^{*F}$,
$\alpha\colon \mathcal{E}(C_{G^*}(s)^*, (1)) \to \mathcal{E}(G,(s))$, such that

$$\text{(8.15)} \qquad \dim \alpha(\rho) = \frac{|G^{*F}|_{p'}}{|C_{G^*}(s)^F|_{p'}} (\dim \rho).$$

<u>Remark</u>. This theorem can be regarded as giving a "Jordan decomposition" for representations of G^F in the sense that each irreducible representation of G^F is associated with a semisimple element s of G^{*F} and a unipotent representation of $C_{G^*}(s)^{*F}$. Lusztig has conjectured that such a theorem holds in general for all G and not just for classical groups.

At this point we prove a result about dimensions of unipotent representations which will be used later. By (6.23), if ρ denotes the regular representation of G^F we have

$$\rho = \frac{1}{|G^F|_p} \sum_T \sum_{\theta \in \hat{T}} \varepsilon_G \varepsilon_T R^G_T(\theta), \quad \text{and hence}$$

$$\sum_{\substack{\phi \in \mathcal{E}(G) \\ \phi \text{ unipotent}}} (\dim \phi)\phi + \sum_{\substack{\phi \in \mathcal{E}(G) \\ \phi \text{ non-unipotent}}} (\dim \phi)\phi = \frac{1}{|G^F|_p}\Big\{\sum_T \varepsilon_T \varepsilon_G R^G_T(1) + \sum_{\substack{T,\theta \\ \theta \neq 1}} \varepsilon_T \varepsilon_G R^G_T(\theta)\Big\}.$$

Hence, using (6.13) we have

$$\sum_{\phi \text{ unipotent}} (\dim \phi)\phi = \frac{1}{|G^F|_p} \sum_T \varepsilon_T \varepsilon_G R^G_T(1).$$

Using the dimension formula (6.21) we then get

$$\sum_{\phi \text{ unipotent}} (\dim \phi)^2 = \frac{1}{|G^F|_p} \sum_T \frac{|G^F|_{p'}}{|T^F|}$$

$$= \frac{1}{|G^F|_p} \sum_{(T)} \frac{|G^F|}{|T^F||W(T)^F|} \cdot \frac{|G^F|_{p'}}{|T^F|} .$$

Thus, using (2.10) we get

(8.16) $$\sum_{\phi \text{ unipotent}} (\dim \phi)^2 = \frac{1}{|W|} \sum_{w \in W} \frac{|G^F|_{p'}^2}{|\det(wF - 1)|^2} .$$

The rest of this section will be devoted to a discussion of the proofs of Theorems 8.12 and 8.14.

Consider a classical Weyl group, i.e., of type A, B or D. Denote by $\mathcal{H}_n(q,y)$ the algebra $\mathcal{H}(W_n,\phi)$ where W_n is the Weyl group of type B_n and $\phi(w_j) = q$ $(1 \leq j \leq n-1)$, $\phi(w_n) = y$ for some y in $\overline{\mathbb{Q}}_\ell$, $w_1, w_2, \ldots, w_n$ being the fundamental reflections, i.e., the elements of S, where w_n corresponds to the short root. Let $\tilde{\mathcal{H}}_n(q)$ be the algebra $\mathcal{H}(\tilde{W}_n,\phi)$ where $\tilde{W}_n$ is of type D_n and $\phi(s) = q$ for all s. Hoefsmit [35] has described the representations of $\mathcal{H}_n(q,y)$ and $\tilde{\mathcal{H}}_n(q)$. Lusztig gives an alternative description of the representations in terms of symbols, rather than pairs of partitions, and proves the following proposition.

<u>Proposition 8.17</u> ([47], 4.6.2). There exists a bijection between the set $\mathcal{H}_n(q,q^d)^\vee$ of irreducible representations over $\overline{\mathbb{Q}}_\ell$ of $\mathcal{H}_n(q,q^d)$ and the set $\Phi_{n',d}$ of equivalence classes of symbols of rank n' and defect d, where $n' = n + [(\frac{d}{2})^2]$. Sup-

pose there is a special (see 8.11) homomorphism $\mathcal{H}_n(q,q^d) \to \mathrm{End}\ E$, where E is a finite-dimensional $\overline{\mathbb{Q}}_\ell$-vector space. If $E_0 \in \mathcal{H}_n(q,q^d)^\vee$ corresponds to $(\Lambda) \in \Phi_{n',d'}$ the multiplicity of E_0 in the $\mathcal{H}_n(q,q^d)$-module E is given by

$(\dim E)\ D_\Lambda(q)\ D_{([0,d-1],\emptyset)}(q)^{-1}(q-1)^n$. [Here $(0,d-1],\emptyset)$ stands for the symbol whose two rows are $\{d-1 > 0\}$ and $\emptyset$.]

A similar proposition ([47], 4.7.1) holds for the algebra $\mathcal{H}_n(q)$.

We will mainly describe the easiest case of Theorems 8.12 and 8.14, i.e. $G = Sp(2n,q)$, $G^* = SO(2n+1,q)$, $n \geq 1$, q even. Then we wish to show the following.

(1) There is a bijection $(\Lambda) \leftrightarrow \rho_\Lambda$ between $\Phi_n = \coprod_{d\ \mathrm{odd}} \Phi_{n,d}$ and $\mathcal{E}(G,(1))$ and the dimension of ρ_Λ is given by (8.13).

(2) If (s) is a semisimple class of G^*, there is a bijection α: $\mathcal{E}(C_{G^*}(s)^*,(1)) \leftrightarrow \mathcal{E}(G,(s))$ such that $\dim \alpha(\rho)$ is given by (8.15).

We first check that the theorem is true for $n = 1$ and assume that it is true for $n' < n$. Now let $s \in G^{*F}$, where $s \neq 1$ is semisimple. We choose a regular subgroup L' of G^* such that $C_{G^*}(s) \subset L'$. (This can always be done in the case when q is even; in classical groups the only semisimple elements for which this fails to hold are elements with eigenvalues ± 1 in the natural representation.) Then there is a regular subgroup L of G corresponding to L'. To prove (2) it is then enough to show that the conjugacy class of s in L'^F has the property (2) with respect to L', i.e., that we have a bijection $\mathcal{E}(C_{L'}(s)^*,(1)) \to \mathcal{E}(L,(s))$. For, composing this with the bijection $\mathcal{E}(L,(s)) \leftrightarrow \mathcal{E}(G,(s))$ obtained from the map $\rho \to \varepsilon_G \varepsilon_L R_L^G(\rho)$ $(\rho \in \mathcal{E}(L,(s))$ and noting that $C_{L'}(s) \cong C_{G^*}(s)$, we have (2). If all the semi-simple components of L' (and hence of L) are of type A then we are done, by the description of the representations of $GL(n,q)$ and $U(n,q)$ given earlier. If not, L has a component isomorphic to Sp_{2m} for some $m < n$ and we can use the induction hypothesis. Thus (2) holds.

We now consider (1). We make the following definitions:

<u>Definition</u> (i) $B(G^F)$ is the number of conjugacy classes of G^F.

(ii) $B_1(G^F)$ is the number of unipotent conjugacy classes of G^F.

We show that in fact $B_1(G^F) = \mathcal{E}(G,(1))$. (This is false if q is odd.) Since q is even , we have an isogeny $SO_{2n+1} \to Sp_{2n}$ which leads to a bijection $(s) \leftrightarrow (s')$ between classes of semisimple elements of G^F and G^{*F} such that $C_G(s) \cong C_{G^*}(s')^*$. Thus

$$|\mathcal{E}(G)| = \sum_{(s')} |\mathcal{E}(C_{G^*}(s'), (1))|, \text{ by (2)}$$
$$= \sum_{(s)} |\mathcal{E}(C_G(s),(1))|.$$

Now the groups $C_G(s)^F$ are products of symplectic groups, general linear groups, or unitary groups (possibly over extensions of $\mathbb{F}_q$

in the last two cases). For general linear groups or unitary groups, the number of unipotent representations is equal to the number of unipotent conjugacy classes, as we have seen. Using this fact and induction on n we may assume that $\mathcal{E}(C_G(s),(1)) = B_1(C_G(s)^F)$ whenever $s \neq 1$. But we have $B(G^F) = \sum_{(s)} B_1(C_G(s)^F)$ by the Jordan decomposition for elements of G^F. This shows that $B_1(G^F) = \mathcal{E}(G,(1))$, as claimed. Thus it is sufficient to show that $B_1(G^F) = |\Phi_n|$. Lusztig proves ([ ], 3.4.1) that

$$\sum_{n=0}^{\infty} |\Phi_n| t^n = \prod_{i=1}^{\infty} (1 - t^i)^{-2} \sum_{j=0}^{\infty} t^{j(j+1)}.$$

Then it follows from a formula of Wall [86] for the number of unipotent classes of $Sp(2n,q)$ (q even) and an identity of Andrews [1] that the right hand side of this identity is also $\sum_{n=0}^{\infty} B_1(G^F)t^n$. This proves the first part of (1).

Next we consider the second part of (1) and also when unipotent cuspidal representations occur in this case. The idea is to induce unipotent cuspidal representations from parabolic subgroups, consider their constituents which are non-cuspidal unipotent representations of G^F, and see what unipotent representations remain, if any.

Let $L' \neq G^*$ be a regular subgroup of G^* which is contained in an F-stable parabolic subgroup. Corresponding to L' we have a regular subgroup L of G which is contained in an F-stable parabolic subgroup P of G. Suppose $\mathcal{E}(L,(1))$ contains some cuspidal representation; then no component of L can be of type A since $GL(n,q)$ has no unipotent cuspidal representations. Thus L is isomorphic over $\mathbb{F}_q$ to $S \times G_1$ where S is a torus and $G_1 \cong Sp_{2m}$ for some $m < n$. By the induction hypothesis we must have $m = [(\frac{d}{2})^2]$ for some odd integer d and G_1^F has a unique unipotent cuspidal representation. By combining it with the trivial representation of S we get a unipotent representation ρ of L^F. We then consider $E = \mathrm{Ind}_{P^F}^{G^F}(\tilde{\rho})$ and use (8.17) to study the constituents of E. The discussion following (8.11) shows that the centralizer algebra of E is of the form $\mathcal{H}_{n-m}(q,y)$ for some y. In order to pin down y we make the following induction assumption.

(8.18) If $\mathcal{E}(G,(1))$ contains <u>subcuspidal</u> representations, i.e., representations which are constituents of the induced representations of a <u>maximal</u> parabolic subgroup, then there are exactly two of these and the ratio of their dimensions is q^t where $[(t/2)^2] = [\frac{\dim V''-2}{2}]$. Here V'' is the $\mathbb{F}_q$-space on which G^* acts in its natural representation.

Now suppose $m \leq n-2$. Then P^F is not a maximal parabolic subgroup of G^F and there is a proper subgroup of G^F containing P^F as a maximal parabolic subgroup to which (8.18) applies. The power of q appearing there is then q^d, where $m = [(d/2)^2]$. Then it can be shown ([47], 5.15) that this is exactly the parameter y, i.e., that the centralizer algebra of E is isomorphic to to $\mathcal{H}_{n-m}(q,q^d)$. The constituents of E are in bijection with the elements of $\mathcal{H}_{n-m}(q,q^d)^{\vee}$ and thus with the elements of $\Phi_{n,d}$, by (8.17). Moreover, using (8.17) we can also show that if $(\Lambda) \varepsilon \Phi_{n,d}$ corresponds to a constituent ρ_Λ of E, then $\dim \rho_\Lambda = D_\Lambda(q)|G^F|_{p'}$.

If $m = n-1$, then we know that the centralizer algebra of E is isomorphic to $\mathcal{H}_1(q,y)$ for some y which is not known. But in this case $\mathcal{H}_1(q,y)$ is of dimension 2 and E has two irreducible constituents. These correspond to the two classes of symbols of rank $[(\frac{d}{2})^2]+1$ and defect d, namely, the classes of

$$\begin{Bmatrix} d > d-1 > \cdots > 0 \\ 1 \end{Bmatrix} \text{ and } \begin{Bmatrix} d > d-2 > d-3 > \cdots > 1 > 0 \\ \emptyset \end{Bmatrix}.$$

The sum of $D_\Lambda(q)$ for these two (Λ) is checked to be $D_{\lambda_0}(q)(q-1)^{-1}$ where $\lambda_0 = \begin{Bmatrix} d-1 > d-2 > \cdots > 0 \\ \emptyset \end{Bmatrix}$, and thus we get, in this case

$$(8.19) \qquad \sum_{(\Lambda)\varepsilon\Phi_{n,d}} \dim \rho_\Lambda = \sum_{(\Lambda)} D_\lambda(q)|G^F|_{p'} \quad . \qquad (\text{Here } [(\tfrac{d}{2})^2]=n-1.)$$

Now we have shown that the number of unipotent representations is Φ_n, and $|\Phi_n| = \sum_{[(\frac{d}{2})^2]\le n,\, d \text{ odd}} |\Phi_{n,d}|$. Now all the non-cuspidal unipotent representations have been accounted for and they are in bijection with $\bigcup_{[(\frac{d}{2})^2]\le n-1} \Phi_{n,d}$. Also we have $|\Phi_{n,d}| = 1$ if

$n = [(\frac{d}{2})^2]$ since the only symbol of rank $[(\frac{d}{2})^2]$ and defect d, up to equivalence, is λ_0 defined above. Thus the number of unipotent cuspidal representations is 1 if $n = [(\frac{d}{2})^2]$ for some odd $d \geq 1$ and 0 otherwise. If $n = [(\frac{d}{2})^2]$ we associate a unipotent cuspidal representation ρ_{λ_0} to λ_0. We get a correspondence $(\Lambda) \to \rho_\Lambda$ from Φ_n onto $\mathcal{E}(G,(1))$. It remains to prove the dimension formula for cuspidal and subcuspidal ρ_Λ, and to prove (8.18) for G.

Now we consider the formula (8.16) for $\sum_{(\Lambda)\in\Phi_n} (\dim \rho_\Lambda)^2$.

It can be shown ([47], 3.5) that $\sum_{(\Lambda)\in\Phi_n} D_\Lambda(q)^2 = \frac{1}{|W_n|} \sum_{w\in W_n} \frac{1}{|\det(wF-1)|^2}$, and thus we have

(8.20) $$\sum_{(\Lambda)\in\Phi_n} (\dim \rho_\Lambda)^2 = \sum_{(\Lambda)\in\Phi_n} D_\lambda(q)^2 |G^F|_{p'}^2 .$$

If $n = [(\frac{d}{2})^2]$ there are no subcuspidal representations since both n and $n - 1$ cannot be of the form $t^2 + t$. Similarly if $n - 1 = [(\frac{d}{2})^2]$ there are no cuspidal representations. So from (8.19) and (8.20) and the fact that we have (8.13) if ρ_Λ is not cuspidal or subcuspidal, we get one equation for $\dim \rho_\Lambda$ if ρ_Λ is cuspidal or two equations for the dimensions of the two subcuspidal ρ_Λ. In each case we find (8.13) holds. Furthermore in the subcuspidal case the ratio of the two dimensions is q^d and thus (8.18) also holds. Thus we have the theorems in Case (a).

Case (a) is the simplest of all the cases, and although many of the arguments are similar in the other cases, some new difficulties appear when q is odd. We will briefly mention these.

A semisimple element $s \in G^{*F}$ is said to be <u>exceptional</u> if its centralizer $C_{G^*}(s)$ has the same semisimple rank as G^*. On V'' (this is the $\mathbb{F}_q$-space on which G^* acts) an element s of this form has eigenvalues 1 or -1. For example, if $G^* = Sp_{2n}$, SO_{2n}, SO_{2n} or SO_{2n+1}, and $s = \begin{pmatrix} 1 & & & & & \\ & 1 & & & & \\ & & \ddots & & & \\ & & & 1 & & \\ & & & & -1 & \\ & & & & & -1 \\ & & & & & & \ddots \\ & & & & & & & -1 \end{pmatrix}$ then the connected centralizer of s in G^* is a product of two symplectic groups in the first case and a product of two orthogonal groups in the other two cases. Such a centralizer cannot be embedded inside a regular subgroup L', and this is a source of difficulties in constructing the representations of G in $\mathcal{E}(G,(s))$.

<u>Example</u>. In the character table of $Sp(4,q)$, q odd (see [76]), the characters denoted by $\Phi_9, \theta_1, \theta_2, \theta_3, \theta_4$ correspond to an exceptional semisimple element in the dual group.

Suppose, for example, we are in Case (c) of (8.14) and let s be a semisimple element of G^{*F} whose only eigenvalues on V'' are ± 1. In this case we can still embed s in a proper regular subgroup L' of G^* which is contained in an F-stable parabolic subgroup, but we might not have $C_{G^*}(s) \subset L'$. We have a corresponding regular subgroup L of G which is contained in an F-stable parabolic subgroup P of G. Suppose $\mathcal{E}(L,(s))$ contains a cuspidal representation. Then L is again of the form $S \times G_1$ where S is a torus and G_1 is a special orthogonal group. Then $\mathcal{E}(L,(s))$ will contain a unique cuspidal representation ρ and as before we would like to decompose $E = \operatorname{Ind}_{P^F}^{G^F}(\tilde{\rho})$.

Now L' is of the form $S' \times G_1'$ where S' is a torus and G_1' is a symplectic group. So we have an orthogonal decomposition of V'' as $V'' = V_0 \oplus \tilde{V}''$ where G_1' is the group of isometries of $\tilde{V}''$. Let $\tilde{V}_1'', \tilde{V}_{-1}''$ be the (± 1)-eigenspaces of s on $\tilde{V}''$. Using the fact that $\mathcal{E}(L,(s))$ has a cuspidal representation which corresponds to a cuspidal representation in $\mathcal{E}(C_L,(s)^*,(1))$ we see that there exist odd integers d, $d-\geq 1$ such that $\frac{1}{2} \dim \tilde{V}_1'' = [(\frac{d}{2})^2]$, $\frac{1}{2} \dim \tilde{V}_{-1}'' = [(\frac{d-}{2})^2]$. Let $\tau = [\frac{1}{2} \dim V_1'']$, $\tau- = [\frac{1}{2} \dim V_{-1}'']$ where V_1'', V_{-1}'' are the 1 and (-1)-eigenspaces of s on V''. Then it can be shown that the centralizer algebra of E is isomorphic to $\mathcal{H}_\ell(q,y) \otimes \mathcal{H}_m(q,y')$ for some y,y' where $\ell = \tau - [(\frac{d}{2})^2]$, $m = (\tau-) - [(\frac{d}{2})^2]$. After this, the analysis proceeds as in Case (a).

Remark. We have seen that in two cases, when $E = \operatorname{Ind}_{B_0^F}^{G^F}(1)$ or $E = \operatorname{Ind}_{P^F}^{G^F}(\tilde{\rho})$ where ρ is a unipotent cuspidal representation, $\operatorname{End} E \cong \mathcal{H}(W,\phi)$ for a suitable W and ϕ. If ρ is an arbitrary cuspidal representation of L^F and $E = \mathrm{Ind}_{P^F}^{G^F}(\tilde{\rho})$, Springer (see [7], C-12) conjectured that End E is isomorphic to the group algebra of the stabilizer of ρ in $N(L)^F/L^F$, twisted by a certain 2-cocycle. Recently R. B. Howlett and G. L. Lehrer have announced a proof of this conjecture. They show that the cocycle is trivial if G has a connected center.

Exceptional groups ([45], [49], [48]).

In the classical groups we have seen that there is at most one unipotent cuspidal representation in each case. It is this miraculous fact that enables us to classify them using the dimension equation (8.16) once the non-cuspidal unipotent representations are known. However, this is not true in the case of the exceptional groups and the number of unipotent cuspidal representations in each case is as follows. We also include some twisted groups here.

Type	Number of unipotent cuspidal representations	
G_2	4	
F_4	7	(q large)
E_6	2	
E_7	2	
E_8	13	(q large)
${}^3D_4$	2	(q large)
${}^2E_6$	3	(q large)
${}^2B_2$	2	
${}^2G_2$	4	
${}^2F_4$	10	(q large)

In the case of the exceptional groups Lusztig used another technique to construct unipotent cuspidal representations (see [45]). First we describe, for any G, an alternative way of realizing the virtual representations $R_T^G(1)$ ([18], 1.4). We say two Borel subgroups B_1, B_2 of G are in relative position w, for some $w \in W(T_0) = W$, if $B_1 = gB_0g^{-1}$, $B_2 = g\dot{w}B_0\dot{w}^{-1}g^{-1}$ for some $g \in G$. Let X_w be the scheme of all Borel subgroups B such that B and FB are in relative position w. Then G^F acts on X_w by conjugation and hence on $H_c^i(X_w)$. The virtual representation of G^F on $\sum_i (-1)^i H_c^i(X_w)$ is the representation R_w. So the idea is to try and decompose the $H_c^i(X_w)$ into irreducible constituents. Now F^δ acts on these spaces, where δ is the smallest integer such that F^δ acts trivially on W (thus $\delta = 1$ if G is split over $\mathbb{F}_q$). One can thus consider the eigenspaces of F^δ on $H_c^i(X_w)$.

<u>Definition</u>. The element $f \in W$ is a <u>Coxeter element</u> if $f = s_1s_2 \cdots s_r$ where $r = \ell(f)$, each $s_i \in S$ (a set of fundamental reflections in W) and $\{s_i\}$ is a set of representatives for the orbits of F on S. The order of the F-centralizer of f is denoted by h_0.

<u>Theorem 8.21</u> ([45], 6.1). Suppose the Dynkin diagram of G is connected. The action of F^δ on $\bigoplus_i H_c^i(X_f)$ is semisimple

and F^δ has precisely h_0 distinct eigenvalues λ_i $(i=1,2,\dots,h_0)$ with multiplicities ≥ 1. The λ_i-eigenspaces are mutually non-isomorphic irreducible G^F-modules.

Thus one realizes unipotent representations of G^F on the eigenspaces of F^δ. The cases where cuspidal representations occur are listed in [48], Table I, p. 32. The eigenvalues of F^δ are products of a root of unity and a power of q^δ. This power is of the form $q^{n\delta}$ where n is an integer except when G is of type E_7 or E_8, when n could be a half-integer.

Remark. In this theorem only the constituents of $\bigoplus_i H_c^i(X_f)$ are considered. In [48], 3.9 it is shown that if ρ is any unipotent representation of G^F, it occurs as a constituent of the generalized μ-eigenspace of F^δ on some $H_c^i(X_w)$ and μ is uniquely determined by ρ up to a factor $q^{n\delta}$ where n is an integer. (See also [49], §1,2.)

If G is of type E_6 or E_7 the two cuspidal unipotent representations of G^F occur as constituents of $H_c^i(X_f)$ as described in (8.21). (See [48], 3.27.) In the case of G_2 $(p\neq 2,3)$ the unipotent cuspidal representations (or rather, their characters) are the ones listed as X_{17}, X_{18}, X_{19}, $\overline{X}_{19}$ by Chang and Ree [10]. If G is of type F_4, there are 7 unipotent cuspidal representations and they are classified in ([48], 3.29). Finally if G is of type E_8, the 13 unipotent cuspidal representations are described in [49].

The dimensions of the unipotent representations are known

in all cases and they are polynomials in q with rational coefficients. There is an interesting pattern to the denominators which occur in these rational coefficients, which has led Lusztig to propose a classification of unipotent representations into families, each family being associated with a certain finite group. In the case of classical groups these finite groups are elementary abelian 2-groups and in the case of exceptional groups they are the symmetric groups S_1, S_2, S_3, S_4 or S_5. For details the reader is referred to [49].

REFERENCES

1. G.E. Andrews, Partitions, q-functions and the Lusztig-Macdonald-Wall conjectures, Inventiones Math. 41 (1977) 91-102.

2. M. Artin, Grothendieck topologies, Harvard Math. Dept. Lecture Notes (1962).

3. C.T. Benson, The generic degrees of the irreducible characters of E_8, to appear in Comm. in Algebra.

4. C.T. Benson, L.C. Grove, D.B. Surowski, Semilinear automorphisms and dimension functions for certain characters of finite Chevalley groups, Math. Z. 144 (1977), 149-159.

5. A. Borel, Linear Algebraic Groups, W.A. Benjamin, Inc. New York, 1969.

6. A. Borel, Seminar on transformation groups, Annals of Math. Studies 46, Princeton Univer. Press, Princeton, N.J. 1960.

7. A. Borel, et al, Seminar on algebraic groups and related finite groups, Lecture Notes in Math. 131, Springer Verlag, 1970.

8. N. Bourbaki, Groupes et algebres de Lie, chap. 4, 5, 6, Act. Sci. Ind., no. 1337, Hermann, Paris, 1968.

9. R.W. Carter, Simple groups of Lie type, John Wiley and Sons, London, 1972.

10. B. Chang and R. Ree, The characters of $G_2(q)$, Symp. Math., vol. XIII (1974), 395-413.

11. C. Chevalley, Classification des groupes de Lie algebriques, Paris 1956/58.

12. C.W. Curtis, Reduction theorems for characters of finite groups of Lie type, J. Math. Soc. Japan 27 (1975), 666-688.

13. __________, A note on the connections between the generalized characters of Deligne and Lusztig and the cuspidal characters of reductive groups over finite fields, J. London Math. Soc. (2) 14 (1976), 405-412.

14. __________, Representations of Finite Groups of Lie type, to appear in the Bull. Amer. Soc.

15. __________, On the values of certain irreducible characters of finite Chevalley groups, Instituto Nazionale di Alta Matematica, Symp. Math. vol. XIII (1974), 343-355.

16. C.W. Curtis, N. Iwahori, R. Kilmoyer, Hecke algebras and characters of parabolic type of finite groups with (B,N)-pairs, Publ. Math. I.H.E.S. 40 (1972), 81-116.

17. C.W. Curtis, and I. Reiner, Representation Theory of Finite Groups and Associative Algebras, John Wiley, 1962.

18. P. Deligne and G. Lusztig, Representations of reductive groups over finite fields, Ann. of Math. 103 (1976), 103-161.

19. M. Demazure and P. Gabriel, Groupes Algebriques, Tome I, North-Holland, 1970.

20. J. Dixmier, Enveloping Algebras, North-Holland, 1977.

21. H. Enomoto, The characters of the finite Chevalley groups $G_2(q)$, $q = 3^b$, Jap. J. Math. (N.S.) 2 (1976), 191-248.

22. ___________, The characters of the finite symplectic group $Sp(4,q)$, $q = 2^b$, Osaka J. Math. 9 (1972), 75-94.

23. V. Ennola, On the characters of the finite unitary groups, Ann. Acad. Sci. Fenn. 323 (1963), 1-35.

24. C. Faith, Algebra: Rings, Modules and Categories I, Springer-Verlag, 1973.

25. J.S. Frame and W. Simpson, The character tables for $SL(3,q)$, $SU(3,q^2)$, $PSL(3,q)$, $PSU(3,q^2)$, Can. J. Math. 25 (1973), 486-494.

26. W. Feit, Characters of finite groups, W.A. Benjamin, Inc., New York, 1967.

27. G. Frobenius, Uber die Charaktere der symmetrische Gruppe, Preuss. Akad. Wiss. Sitzungsber. (1900), 516-534.

28. A. Grothendieck, Formula de Lefschetz et rationalite des fonctions-L, Sem. Bourbaki 279, W.A. Benjamin, New York, 1966.

29. R. Godement, Topologie Algebrique et Theorie des Faisceaux, Hermann, Paris, 1958.

30. J.A. Green, The characters of the finite general linear groups, Trans. Amer. Math Soc. 80 (1955), 402-447.

31. P.J. Hilton and U. Stammbach, A course in Homological Algebra, Graduate Texts in Mathematics 4. Springer-Verlag, 1970.

32. R. Hartshorne, Algebraic Geometry, Graduate Texts in Mathematics 52, Springer-Verlag, 1970.

33. R. Hartshorne, Residues and Duality, Lecture Notes in Math. 20, Springer-Verlag, 1966.

34. G. Hochshild, The structure of Lie groups, Holden-Day, Inc., San Francisco, 1965.

35. P.N. Hoefsmit, Preprsentations of Hecke algebras of finite groups with BN pairs of classical type, Thesis, Univ. of British Columbia, Vancouver, 1974.

36. R. Hotta and K Matsui, On a lemma of Tate-Thompson, Hiroshima Math. J. 8 (1978), 255-268.

37. R. Hotta and T.A. Springer, A specialization theorem for certain Weyl group representations and an application to the Green polynomials of unitary groups, Invent. Math. 41 (1977), 113-127.

38. R.B. Howlett and R.W. Kilmoyer, Principal series representations of finite groups with split BN-pairs, to appear in Comm. in Algebra.

39. J.E. Humphreys, Introduction to Lie Algebras and Representation Theory, Graduate Texts in Mathematics 9, Springer-Verlag, 1972.

40. ____________, Linear Algebraic Groups, Graduate Texts in Mathematics 21, Springer-Verlag, 1975.

41. D. Kazhdan, Proof of Springer's Hypothesis, Israel J. Math. 28 (1977), 272-286.

42. R.W. Kilmoyer, Principal series representations of finite Chevalley groups, J. Algebra 51 (1978), 300-319.

43. R.L. Lipsman, Group Representations, Lecture Notes in Mathematics 388, Springer-Verlag, 1974.

44. G. Lusztig, On the finiteness of the number of unipotent classes, Inventiones Math. 34 (1976), 201-213. MR 54, #7653.

45. _________, Coxeter orbits and eigenspaces of Frobenius, Inventiones Math. 28 (1975), 101-159; MR 56, #12138.

46. _________, On the Green polynomials of classical groups, Proc. London Math. Soc. (3) 33 (1976), 443-475.

47. _________, Irreducible representations of finite classical groups, Inventiones Math. 43 (1977), 125-175; MR 57, #3228.

48. __________, Representations of finite Chevalley groups, C.B.M.S. Regional Conference Series in Mathematics 39, American Mathematical Society, Providence, Rhode Island.

49. __________, Unipotent representations of a finite Chevalley group of type E_8, to appear in Quarterly J. Math.

50. G. Lusztig and B. Srinivasan, The characters of the finite unitary groups, J. of Algebra 49 (1977), 167-171; MR 56 #12139.

51. I.G. Macdonald, Algebraic Geometry: Introduction to Schemes, W.A. Benjamin, In., New York, 1968.

52. A.O. Morris, A survey on Hall-Littlewood functions and their applications to Representation Theory, in Combinatoire et Representation du Groupe Symetrique, Lecture Notes in Math. 579, Springer-Verlag, 1977.

53. D. Mumford, Picard groups of moduli problems, in Arithmetical Algebraic Geometry (Schilling, ed.) Harper & Row, New York, 1965.

54. __________, Introduction to Algebraic Geometry, Harvard Math. Dept. Lecture Notes.

55. __________, Abelian Varieties, Oxford University Press, Oxford, 1970.

56. J.P. Murre, Lectures on an introduction to Grothendieck's Theory of the Fundamental Group, Lecture Notes,Tata Institute of Fundamental Research, Bombay, 1967.

57. S. Nozawa, On the characters of the finite general unitary group $U(4,q^2)$ Journal of the Fac. Science, Univ. of Tokyo, 19 (1972), 257-293; On the characters of the finite general unitary group $U(5,q^2)$, ibid. 23 (1976), 23-74.

58. R. Ree, On some simple groups defined by C. Chevalley, Trans. Amer. Math. Soc. 84 (1957), 392-400.

59. M. Rosenlicht, A Remark on Quotient Spaces, An. Acad. Brasil Ci. 35 (1963), 487-489; MR 30 #2009.

60. SGA 1: Revetements etales et groupe fondamental dirigé par A. Grothendieck), Lecture Notes in Math. 244, Springer-Verlag, 1971.

61. SGA 3: Schemas en Groupes (dirigé par A Grothendieck et M. Demazure) Lecture Notes in Math. 151, 152, 153, Springer-Verlag, 1970.

62. SGA 4: Theorie des topos et cohomologie etale des schemas (dirigé par M.Artin, A. Grothendieck, et J.L. Verdier) Lecture Notes in Math. 269, 270, 305, Springer-Verlag, 1972/73.

63. SGA 4 1/2: Cohomolgie Etale (dirigé par P. Deligne), L Lecture Notes in Math. 569, Springer-Verlag, 1977.

64. SGA 5: Cohomologie ℓ-adique et Fonctions L (dirigé par A. Grothendieck), Lect re Notes in Math. 589, Springer-Verlag, 1977.

65. SGA 7: Groupes de Monodsomie et Theoreme de Riemann-Roch, I (par A. Grothendieck), II (par P. Deligne et N. Katz), Springer Lecture Notes 288, 340, Springer-Verlag, 1972/73.

66. J.-P. Serre, Linear Representations of Finite Groups, Graduate Texts in Mathematics 42, Springer, Verlag, 1977.

67. __________, Groupes algebriques et corps de classes, Hermann, Paris, 1959.

68. __________, Valeurs propres des endomorphismes de Frobenius, Sem. Bourbaki 446, 1973/74, Lecture Notes in Math. 431, Springer-Verlag 1975.

69. __________, Representations lineaires des groupes finis "algebriques", Sem. Bourbaki 487, 1975/76, Lecture Notes in Math. 567, Springer-Verlag.

70. T. Shoji, On the Springer representations of the Weyl groups of classical groups, to appear in Comm. in Algebra; On the Springer representations of Chevalley groups of type F_4, Preprint.

71. N. Spaltenstein, The fixes point set of a unipotent transformation on the flag manifold, Proc. Kon. Ak. v. Wet. (Amsterdam) 79 (1976), 452-456.

72. T.A. Springer, On the characters of finite groups, in Lie groups and their representations, Proc. Summer School, Budapest, 1975.

73. __________, Caracteres de groupes de Chevalley finis, Sem. Bourbaki 429, 1972/73, Lecture Notes in Math. 383, Springer-Verlag.

74. __________, Generalization of Green's polynomials, Proc. Symp. Pure Math. XXI (1971), Amer. Math. Soc., 149-153.

75. ______________, Trigonometric sums, Green functions of finite groups and representations of Weyl groups, Inventiones Math. 36 (1976), 173-207; MR 56, #491.

76. B. Srinivasan, The characters of the finite symplectic group Sp(4,q), Trans. Amer. Math. Soc. 131 (1968), 488-525.

77. ______________, Green polynomials of finite classical groups Comm. in Alg. 5 (1977), 1241-1258.

78. R. Steinberg, A Geometric approach to the representations of the full linear group over a Galois field , Trans. Amer. Math. Soc. 71 (1951), 274-282.

79. ______________, The representations of GL(3,q), GL(4,q), PGL(3,q), and PGL(4,q), Canadian J. Math. 3 (1951), 225-235.

80. ______________, Conjugacy classes in Algebraic Groups, Lecture Notes in Math. 366, Springer-Verlag, 1974.

81. ______________, Lectures on Chevalley Groups, Yale Univ., Math. Dept. Lecture Notes, 1968.

82. ______________, Endomorphisms of linear algebraic groups, Mem. Amer. Math. Soc. 80 (1968).

83. ______________, On the Desingularization of the Unipotent Variety, Inventiones Math. 36 (1976), 209-224.

84. ______________, On theorems of Lie-Kolchin, Borel and Lang, pp. 349-354 in "Contributions to Algebra: A Collection of Papers dedicated to Ellis Kolchin", Academic Press, 1977.

85. J. Tate, Algebraic cycles and poles of the zeta function, in Arithmetical Algebraic Geometry (Schilling, ed.), Harper & Row, 1965, 93-110.

86. G.E. Wall, On the conjugacy classes in the unitary, symplectic and orthogonal groups, J. Austral. Math. Soc. 3 (1963), 1-62.

87. H.N. Ward, On Ree's series of simple groups, Trans. Amer. Math. Soc. 121 (1966), 62-89.

NOTATION

$\mathbb{Z}$: Integers

$\mathbb{Q}$: Rational numbers

$\mathbb{R}$: Real numbers

$\mathbb{C}$: Complex numbers

$\mathbb{F}_q$: Field of q elements, of characteristic p.

Z_ℓ, Q_ℓ: ℓ-adic integers, ℓ-adic numbers

If k is a field, $\bar{k}$ is its algebraic closure

Is X is a subset of a group H and $g \in H$, $X^g = g^{-1}Xg$

and ${}^gX = gXg^{-1}$

Let H be a finite group. Then $|H|_p$ is the p-part of $|H|$,

and $|H|_{p'}$ is the part of $|H|$ prime to p.

$\mathcal{R}(H)$ is the Grothendieck group of H.

If L is a subgroup of H, and ρ is a representation of L,

$\mathrm{Ind}_L^H(\rho)$ is the representation of H induced by ρ.

Index

Vol. 609: General Topology and Its Relations to Modern Analysis and Algebra IV. Proceedings 1976. Edited by J. Novák. XVIII, 225 pages. 1977.

Vol. 610: G. Jensen, Higher Order Contact of Submanifolds of Homogeneous Spaces. XII, 154 pages. 1977.

Vol. 611: M. Makkai and G. E. Reyes, First Order Categorical Logic. VIII, 301 pages. 1977.

Vol. 612: E. M. Kleinberg, Infinitary Combinatorics and the Axiom of Determinateness. VIII, 150 pages. 1977.

Vol. 613: E. Behrends et al., L^P-Structure in Real Banach Spaces. X, 108 pages. 1977.

Vol. 614: H. Yanagihara, Theory of Hopf Algebras Attached to Group Schemes. VIII, 308 pages. 1977.

Vol. 615: Turbulence Seminar, Proceedings 1976/77. Edited by P. Bernard and T. Ratiu. VI, 155 pages. 1977.

Vol. 616: Abelian Group Theory, 2nd New Mexico State University Conference, 1976. Proceedings. Edited by D. Arnold, R. Hunter and E. Walker. X, 423 pages. 1977.

Vol. 617: K. J. Devlin, The Axiom of Constructibility: A Guide for the Mathematician. VIII, 96 pages. 1977.

Vol. 618: I. I. Hirschman, Jr. and D. E. Hughes, Extreme Eigen Values of Toeplitz Operators. VI, 145 pages. 1977.

Vol. 619: Set Theory and Hierarchy Theory V, Bierutowice 1976. Edited by A. Lachlan, M. Srebrny, and A. Zarach. VIII, 358 pages. 1977.

Vol. 620: H. Popp, Moduli Theory and Classification Theory of Algebraic Varieties. VIII, 189 pages. 1977.

Vol. 621: Kauffman et al., The Deficiency Index Problem. VI, 112 pages. 1977.

Vol. 622: Combinatorial Mathematics V, Melbourne 1976. Proceedings. Edited by C. Little. VIII, 213 pages. 1977.

Vol. 623: I. Erdelyi and R. Lange, Spectral Decompositions on Banach Spaces. VIII, 122 pages. 1977.

Vol. 624: Y. Guivarc'h et al., Marches Aléatoires sur les Groupes de Lie. VIII, 292 pages. 1977.

Vol. 625: J. P. Alexander et al., Odd Order Group Actions and Witt Classification of Innerproducts. IV, 202 pages. 1977.

Vol. 626: Number Theory Day, New York 1976. Proceedings. Edited by M. B. Nathanson. VI, 241 pages. 1977.

Vol. 627: Modular Functions of One Variable VI, Bonn 1976. Proceedings. Edited by J.-P. Serre and D. B. Zagier. VI, 339 pages. 1977.

Vol. 628: H. J. Baues, Obstruction Theory on the Homotopy Classification of Maps. XII, 387 pages. 1977.

Vol. 629: W.A. Coppel, Dichotomies in Stability Theory. VI, 98 pages. 1978.

Vol. 630: Numerical Analysis, Proceedings, Biennial Conference, Dundee 1977. Edited by G. A. Watson. XII, 199 pages. 1978.

Vol. 631: Numerical Treatment of Differential Equations. Proceedings 1976. Edited by R. Bulirsch, R. D. Grigorieff, and J. Schröder. X, 219 pages. 1978.

Vol. 632: J.-F. Boutot, Schéma de Picard Local. X, 165 pages. 1978.

Vol. 633: N. R. Coleff and M. E. Herrera, Les Courants Résiduels Associés à une Forme Méromorphe. X, 211 pages. 1978.

Vol. 634: H. Kurke et al., Die Approximationseigenschaft lokaler Ringe. 204 Seiten. 1978.

Vol. 635: T. Y. Lam, Serre's Conjecture. XVI, 227 pages. 1978.

Vol. 636: Journées de Statistique des Processus Stochastiques, Grenoble 1977, Proceedings. Edité par Didier Dacunha-Castelle et Bernard Van Cutsem. VII, 202 pages. 1978.

Vol. 637: W. B. Jurkat, Meromorphe Differentialgleichungen. VII, 4 Seiten. 1978.

Vol. 638: P. Shanahan, The Atiyah-Singer Index Theorem, An Introduction. V, 224 pages. 1978.

Vol. 639: N. Adasch et al., Topological Vector Spaces. V, 125 pages. 1978.

Vol. 640: J. L. Dupont, Curvature and Characteristic Classes. X, 175 pages. 1978.

Vol. 641: Séminaire d'Algèbre Paul Dubreil, Proceedings Paris 1976–1977. Edité par M. P. Malliavin. IV, 367 pages. 1978.

Vol. 642: Theory and Applications of Graphs, Proceedings, Michigan 1976. Edited by Y. Alavi and D. R. Lick. XIV, 635 pages. 1978.

Vol. 643: M. Davis, Multiaxial Actions on Manifolds. VI, 141 pages. 1978.

Vol. 644: Vector Space Measures and Applications I, Proceedings 1977. Edited by R. M. Aron and S. Dineen. VIII, 451 pages. 1978.

Vol. 645: Vector Space Measures and Applications II, Proceedings 1977. Edited by R. M. Aron and S. Dineen. VIII, 218 pages. 1978.

Vol. 646: O. Tammi, Extremum Problems for Bounded Univalent Functions. VIII, 313 pages. 1978.

Vol. 647: L. J. Ratliff, Jr., Chain Conjectures in Ring Theory. VIII, 133 pages. 1978.

Vol. 648: Nonlinear Partial Differential Equations and Applications, Proceedings, Indiana 1976–1977. Edited by J. M. Chadam. VI, 206 pages. 1978.

Vol. 649: Séminaire de Probabilités XII, Proceedings, Strasbourg, 1976–1977. Edité par C. Dellacherie, P. A. Meyer et M. Weil. VIII, 805 pages. 1978.

Vol. 650: C*-Algebras and Applications to Physics. Proceedings 1977. Edited by H. Araki and R. V. Kadison. V, 192 pages. 1978.

Vol. 651: P. W. Michor, Functors and Categories of Banach Spaces. VI, 99 pages. 1978.

Vol. 652: Differential Topology, Foliations and Gelfand-Fuks-Cohomology, Proceedings 1976. Edited by P. A. Schweitzer. XIV, 252 pages. 1978.

Vol. 653: Locally Interacting Systems and Their Application in Biology. Proceedings, 1976. Edited by R. L. Dobrushin, V. I. Kryukov and A. L. Toom. XI, 202 pages. 1978.

Vol. 654: J. P. Buhler, Icosahedral Golois Representations. III, 143 pages. 1978.

Vol. 655: R. Baeza, Quadratic Forms Over Semilocal Rings. VI, 199 pages. 1978.

Vol. 656: Probability Theory on Vector Spaces. Proceedings, 1977. Edited by A. Weron. VIII, 274 pages. 1978.

Vol. 657: Geometric Applications of Homotopy Theory I, Proceedings 1977. Edited by M. G. Barratt and M. E. Mahowald. VIII, 459 pages. 1978.

Vol. 658: Geometric Applications of Homotopy Theory II, Proceedings 1977. Edited by M. G. Barratt and M. E. Mahowald. VIII, 487 pages. 1978.

Vol. 659: Bruckner, Differentiation of Real Functions. X, 247 pages. 1978.

Vol. 660: Equations aux Dérivée Partielles. Proceedings, 1977. Edité par Pham The Lai. VI, 216 pages. 1978.

Vol. 661: P. T. Johnstone, R. Paré, R. D. Rosebrugh, D. Schumacher, R. J. Wood, and G. C. Wraith, Indexed Categories and Their Applications. VII, 260 pages. 1978.

Vol. 662: Akin, The Metric Theory of Banach Manifolds. XIX, 306 pages. 1978.

Vol. 663: J. F. Berglund, H. D. Junghenn, P. Milnes, Compact Right Topological Semigroups and Generalizations of Almost Periodicity. X, 243 pages. 1978.

Vol. 664: Algebraic and Geometric Topology, Proceedings, 1977. Edited by K. C. Millett. XI, 240 pages. 1978.

Vol. 665: Journées d'Analyse Non Linéaire. Proceedings, 1977. Edité par P. Bénilan et J. Robert. VIII, 256 pages. 1978.

Vol. 666: B. Beauzamy, Espaces d'Interpolation Réels: Topologie et Géometrie. X, 104 pages. 1978.

Vol. 667: J. Gilewicz, Approximants de Padé. XIV, 511 pages. 1978.

Vol. 668: The Structure of Attractors in Dynamical Systems. Proceedings, 1977. Edited by J. C. Martin, N. G. Markley and W. Perrizo. VI, 264 pages. 1978.

Vol. 669: Higher Set Theory. Proceedings, 1977. Edited by G. H. Müller and D. S. Scott. XII, 476 pages. 1978.

Vol. 670: Fonctions de Plusieurs Variables Complexes III, Proceedings, 1977. Edité par F. Norguet. XII, 394 pages. 1978.

Vol. 671: R. T. Smythe and J. C. Wierman, First-Passage Perculation on the Square Lattice. VIII, 196 pages. 1978.

Vol. 672: R. L. Taylor, Stochastic Convergence of Weighted Sums of Random Elements in Linear Spaces. VII, 216 pages. 1978.

Vol. 673: Algebraic Topology, Proceedings 1977. Edited by P. Hoffman, R. Piccinini and D. Sjerve. VI, 278 pages. 1978.

Vol. 674: Z. Fiedorowicz and S. Priddy, Homology of Classical Groups Over Finite Fields and Their Associated Infinite Loop Spaces. VI, 434 pages. 1978.

Vol. 675: J. Galambos and S. Kotz, Characterizations of Probability Distributions. VIII, 169 pages. 1978.

Vol. 676: Differential Geometrical Methods in Mathematical Physics II, Proceedings, 1977. Edited by K. Bleuler, H. R. Petry and A. Reetz. VI, 626 pages. 1978.

Vol. 677: Séminaire Bourbaki, vol. 1976/77, Exposés 489–506. IV, 264 pages. 1978.

Vol. 678: D. Dacunha-Castelle, H. Heyer et B. Roynette. Ecole d'Eté de Probabilités de Saint-Flour. VII-1977. Edité par P. L. Hennequin. IX, 379 pages. 1978.

Vol. 679: Numerical Treatment of Differential Equations in Applications, Proceedings, 1977. Edited by R. Ansorge and W. Törnig. IX, 163 pages. 1978.

Vol. 680: Mathematical Control Theory, Proceedings, 1977. Edited by W. A. Coppel. IX, 257 pages. 1978.

Vol. 681: Séminaire de Théorie du Potentiel Paris, No. 3, Directeurs: M. Brelot, G. Choquet et J. Deny. Rédacteurs: F. Hirsch et G. Mokobodzki. VII, 294 pages. 1978.

Vol. 682: G. D. James, The Representation Theory of the Symmetric Groups. V, 156 pages. 1978.

Vol. 683: Variétés Analytiques Compactes, Proceedings, 1977. Edité par Y. Hervier et A. Hirschowitz. V, 248 pages. 1978.

Vol. 684: E. E. Rosinger, Distributions and Nonlinear Partial Differential Equations. XI, 146 pages. 1978.

Vol. 685: Knot Theory, Proceedings, 1977. Edited by J. C. Hausmann. VII, 311 pages. 1978.

Vol. 686: Combinatorial Mathematics, Proceedings, 1977. Edited by D. A. Holton and J. Seberry. IX, 353 pages. 1978.

Vol. 687: Algebraic Geometry, Proceedings, 1977. Edited by L. D. Olson. V, 244 pages. 1978.

Vol. 688: J. Dydak and J. Segal, Shape Theory. VI, 150 pages. 1978.

Vol. 689: Cabal Seminar 76–77, Proceedings, 1976–77. Edited by A.S. Kechris and Y. N. Moschovakis. V, 282 pages. 1978.

Vol. 690: W. J. J. Rey, Robust Statistical Methods. VI, 128 pages. 1978.

Vol. 691: G. Viennot, Algèbres de Lie Libres et Monoïdes Libres. III, 124 pages. 1978.

Vol. 692: T. Husain and S. M. Khaleelulla, Barrelledness in Topological and Ordered Vector Spaces. IX, 258 pages. 1978.

Vol. 693: Hilbert Space Operators, Proceedings, 1977. Edited by J. M. Bachar Jr. and D. W. Hadwin. VIII, 184 pages. 1978.

Vol. 694: Séminaire Pierre Lelong – Henri Skoda (Analyse) Année 1976/77. VII, 334 pages. 1978.

Vol. 695: Measure Theory Applications to Stochastic Analysis, Proceedings, 1977. Edited by G. Kallianpur and D. Kölzow. XII, 261 pages. 1978.

Vol. 696: P. J. Feinsilver, Special Functions, Probability Semigroups, and Hamiltonian Flows. VI, 112 pages. 1978.

Vol. 697: Topics in Algebra, Proceedings, 1978. Edited by M. F. Newman. XI, 229 pages. 1978.

Vol. 698: E. Grosswald, Bessel Polynomials. XIV, 182 pages. 1978.

Vol. 699: R. E. Greene and H.-H. Wu, Function Theory on Manifolds Which Possess a Pole. III, 215 pages. 1979.

Vol. 700: Module Theory, Proceedings, 1977. Edited by C. Faith and S. Wiegand. X, 239 pages. 1979.

Vol. 701: Functional Analysis Methods in Numerical Analysis, Proceedings, 1977. Edited by M. Zuhair Nashed. VII, 333 pages. 1979.

Vol. 702: Yuri N. Bibikov, Local Theory of Nonlinear Analytic Ordinary Differential Equations. IX, 147 pages. 1979.

Vol. 703: Equadiff IV, Proceedings, 1977. Edited by J. Fábera. XIX, 441 pages. 1979.

Vol. 704: Computing Methods in Applied Sciences and Engineering, 1977, I. Proceedings, 1977. Edited by R. Glowinski and J. L. Lions. VI, 391 pages. 1979.

Vol. 705: O. Forster und K. Knorr, Konstruktion verseller Familien kompakter komplexer Räume. VII, 141 Seiten. 1979.

Vol. 706: Probability Measures on Groups, Proceedings, 1978. Edited by H. Heyer. XIII, 348 pages. 1979.

Vol. 707: R. Zielke, Discontinuous Čebyšev Systems. VI, 111 pages. 1979.

Vol. 708: J. P. Jouanolou, Equations de Pfaff algébriques. V, 255 pages. 1979.

Vol. 709: Probability in Banach Spaces II. Proceedings, 1978. Edited by A. Beck. V, 205 pages. 1979.

Vol. 710: Séminaire Bourbaki vol. 1977/78, Exposés 507–524. IV, 328 pages. 1979.

Vol. 711: Asymptotic Analysis. Edited by F. Verhulst. V, 240 pages. 1979.

Vol. 712: Equations Différentielles et Systèmes de Pfaff dans le Champ Complexe. Edité par R. Gérard et J.-P. Ramis. V, 364 pages. 1979.

Vol. 713: Séminaire de Théorie du Potentiel, Paris No. 4. Edité pa F. Hirsch et G. Mokobodzki. VII, 281 pages. 1979.

Vol. 714: J. Jacod, Calcul Stochastique et Problèmes de Martingale X, 539 pages. 1979.

Vol. 715: Inder Bir S. Passi, Group Rings and Their Augmentatio Ideals. VI, 137 pages. 1979.

Vol. 716: M. A. Scheunert, The Theory of Lie Superalgebras. X, 27 pages. 1979.

Vol. 717: Grosser, Bidualräume und Vervollständigungen von Ba nachmoduln. III, 209 pages. 1979.

Vol. 718: J. Ferrante and C. W. Rackoff, The Computational Com plexity of Logical Theories. X, 243 pages. 1979.

Vol. 719: Categorial Topology, Proceedings, 1978. Edited by H Herrlich and G. Preuß. XII, 420 pages. 1979.

Vol. 720: E. Dubinsky, The Structure of Nuclear Fréchet Spaces V, 187 pages. 1979.

Vol. 721: Séminaire de Probabilités XIII. Proceedings, Strasbourg 1977/78. Edité par C. Dellacherie, P. A. Meyer et M. Weil. VII, 64 pages. 1979.

Vol. 722: Topology of Low-Dimensional Manifolds. Proceedings 1977. Edited by R. Fenn. VI, 154 pages. 1979.

Vol. 723: W. Brandal, Commutative Rings whose Finitely Generate Modules Decompose. II, 116 pages. 1979.

Vol. 724: D. Griffeath, Additive and Cancellative Interacting Partic Systems. V, 108 pages. 1979.

Vol. 725: Algèbres d'Opérateurs. Proceedings, 1978. Edité pa P. de la Harpe. VII, 309 pages. 1979.

Vol. 726: Y.-C. Wong, Schwartz Spaces, Nuclear Spaces and Tens Products. VI, 418 pages. 1979.

Vol. 727: Y. Saito, Spectral Representations for Schrödinger Opera tors With Long-Range Potentials. V, 149 pages. 1979.

Vol. 728: Non-Commutative Harmonic Analysis. Proceedings, 197 Edited by J. Carmona and M. Vergne. V, 244 pages. 1979.